Palgrave Studies in Climate Resilient Societies

Series Editor
Robert C. Brears, Avonhead, New Zealand

The Palgrave Studies in Climate Resilient Societies series provides readers with an understanding of what the terms **resilience and climate resilient** societies mean; the best practices and lessons learnt from various governments, in both non-OECD and OECD countries, implementing climate resilience policies (in other words what is 'desirable' or 'undesirable' when building climate resilient societies); an understanding of what a resilient society potentially looks like; knowledge of when resilience building requires slow transitions or rapid transformations; and knowledge on how governments can create coherent, forward-looking and flexible policy innovations to build climate resilient societies that: support the conservation of ecosystems; promote the sustainable use of natural resources; encourage sustainable practices and management systems; develop resilient and inclusive communities; ensure economic growth; and protect health and livelihoods from climatic extremes.

Yulia Ermolaeva · Robert C. Brears ·
Anna Minakova · Pius Siakwah ·
Obodai Torto

Rethinking Water and Energy for a Sustainable Future

palgrave
macmillan

Yulia Ermolaeva
Institute of Sociology
Federal Centre of Theoretical and Applied
Sociology Russian Academy of Sciences
Moscow, Russia

Anna Minakova
"ESG-Expert" LLC company
Moscow, Russia

Obodai Torto
Institute of African Studies (IAS)
University of Ghana
Accra, Ghana

Robert C. Brears
Our Future Water
Christchurch, Canterbury, New Zealand

Pius Siakwah
Institute of African Studies (IAS)
University of Ghana
Accra, Ghana

ISSN 2523-8124   ISSN 2523-8132   (electronic)
Palgrave Studies in Climate Resilient Societies
ISBN 978-3-032-04484-6   ISBN 978-3-032-04485-3   (eBook)
https://doi.org/10.1007/978-3-032-04485-3

This Palgrave Macmillan imprint is published by the registered company Springer Nature Switzerland
AG
The registered company address is: Gewerbestrasse 11, 6330 Cham, Switzerland

If disposing of this product, please recycle the paper.

Competing Interests The authors have no competing interests to declare that are relevant to the content of this manuscript.

About this Book

This book adopts a multidisciplinary approach to explore the intricate relationship between water, energy, and societal well-being. It combines environmental, economic, social, and technological perspectives to comprehensively analyse water's crucial role in energy generation and its essential function in supporting life. The text highlights the need for integrated strategies to address water management and energy production challenges, presenting a comprehensive overview that encourages a holistic view of these interconnected issues.

The publication provides a nuanced narrative on water's pivotal role in energy security, resource conservation, and sustainable development through evidence-based insights, theoretical frameworks, and global case studies. It aims to inform and inspire a re-evaluation of practices and policies to achieve a balanced coexistence between human activities and natural resources. This book serves as a resource for researchers, policymakers, and practitioners, offering innovative solutions and methodologies for addressing the complexities of the water-energy nexus in the pursuit of sustainability.

Contents

About the Authors

Yulia Ermolaeva is an interdisciplinary sustainability researcher, environmentalist, and specialist in environmental social sciences (sociology, psychology, neuropsychology) and sustainability sciences, with an academic background in these fields. She has implemented over 20 global, international, and national projects. As an author and co-author, she has published more than 90 peer-reviewed scientific articles including chapter books, focusing on various aspects of sustainable development, sustainable sciences and environmental sociology. Since 2010, she has been working at the Institute of Sociology of the Federal Centre of Theoretical and Applied Sociology of the Russian Academy of Sciences. In 2025, she founded the Future Compass Sustainability Research and Coaching, and provided research and coaching services for different stakeholders.

Robert C. Brears is the Founder of Our Future Water and Editor-in-Chief of the *Palgrave Handbook of Climate Resilient Societies and the Palgrave Encyclopedia of Urban and Regional Futures*. He is the author of Financing Water Security and Green Growth (Oxford University Press), Urban Water Security (Wiley), The Green Economy and the Water-Energy-Food Nexus (Palgrave Macmillan), Natural Resource Management and the Circular Economy (Palgrave Macmillan), Blue and Green Cities: The Role of Blue-Green Infrastructure in Managing Urban Water Resources (Palgrave Macmillan), Climate Resilient Water Resources Management (Palgrave Macmillan), Developing the Circular Water Economy (Palgrave Macmillan), Nature-Based Solutions to 21st Century Challenges (Routledge), Regional Water Security (Wiley), and Water Resources Management: Innovative and Green Solutions (De Gruyter). He is the editor of the Climate Resilient Societies book series published with Palgrave Macmillan.

Anna Minakova is the CEO "ESG-Expert" LLC. Certified more than 1,000,000 sq m of Russian office buildings that are involved in green real estate industry and two of business offices were provided as the best case studies on BRE Global website. In 2021 the first in Russia Fitwell Ambassador for Alfa Bank and Space1 offices. Alfa Bank became the winner of Best in Building Health Winners in the nomination Highest Scoring Projects. Cofounder of green building system Clever for New Construction and In-Use. Lecturer at the British Higher School of Design.

Pius Siakwah (Ph.D.) Senior Research Fellow at the University of Ghana, Institute of African Studies. Pius is a resource geographer, teaching and research interests include political economy, natural resources governance, sustainable development, renewable energy transitions, climate change smart agriculture, precarious work, and tourism. Pius is well published in geography, natural resources, development studies, energy, and tourism. Pius holds a Ph.D. in Geography from Trinity College Dublin (TCD), Ireland.

Dr. Obodai Torto is a development sociologist and currently works as a Research Fellow in the Institute of African Studies, University of Ghana, where he is an affiliate lecturer in the Center for Migration Studies and Center for Social Policy Studies. His research focuses on political economy of development, peacebuilding, security, and humanitarian aid, globalization, energy transition, natural resources management and development, sustainability, rural & agricultural development, and forced migration/displacement. Obodai's areas of Ph.D. and MA supervision include foreign aid and development, refugees and humanitarian aid, regional integration, migration-conflict nexus, agricultural development, human trafficking and smuggling, energy regulatory policy, and mining sector. He received his Ph.D. in Sociology (Development Studies option) at the University of Waterloo, Canada.

List of Figures

List of Tables

1

Introduction: Role of Water and Energy in Human Civilization

Yulia Ermolaeva

Abstract This chapter will delve into the foundational role of water and energy throughout human history, underscoring their profound influence on the development and evolution of societies with tracing the utilization of these essential resources from ancient times to the present, the chapter aims to illuminate the intrinsic link between natural resources, cultural evolution, and societal advancement. Here will be covered historical and cultural foundations of exploring water and energy's deep-rooted significance in shaping human civilizations, focusing on technological evolution and societal impact: An examination of how the availability and management of water and energy resources influenced the structure and dynamics of early human settlements and actors acitivity, fostering societal growth and hierarchical structures with a reflection on how water and energy have not only shaped human life materially but also influenced societal values and ethical considerations, setting the stage for future shifts in global energy paradigms.

Keywords Cross-cultural observation · Water management · Energy · Environmental sociology · Value creation · Environmental

© The Author(s), under exclusive license to Springer Nature Switzerland AG 2025
Y. Ermolaeva et al., *Rethinking Water and Energy for a Sustainable Future*, Palgrave Studies in Climate Resilient Societies, https://doi.org/10.1007/978-3-032-04485-3_1

Table 1.1 Key epochs and value shifts

Period	Energy-water nexus	Dominant values
3000 BCE	Riverine irrigation	Communal reciprocity
1750–1900	Hydropower/coal	Productivity, capitalism
1950–Present	Mega-dams, fossil fuels	Risk management, environmental rights
21st Century	Renewables, desalination	Sustainability, equity

modernization · Environmental history · Societal transformations · Subject–object relationships

The interplay between water, energy, and human societies is a narrative etched into the bedrock of civilization. To decode this relationship, we must transcend disciplinary boundaries, studying together theories from anthropology, political ecology, economics, and environmental history. Water became a sociotechnical assemblage—a force that shapes power structures, cultural identities, and technological innovation. Early human settlements coalesced around rivers, transforming water into a social, cultural and strategic asset. The **hydraulic hypothesis** (Wittfogel, 1957) posits that large-scale irrigation systems necessitated centralized authority, birthing "hydraulic civilizations" like Mesopotamia and Egypt, Venetian and Rome Empire. The interplay between water, energy, and societal values is a dynamic dialectic sunject and object socio technical relationships, where material practices shape moral frameworks and vice versa. We discuss how water and energy systems have molded ethical and environmental norms across epochs—from communal agrarian rites to industrial commodification and modern sustainability imperatives. However, this deterministic view is critiqued by **resilience** (Isendahl & Scarborough, 2004), which emphasizes adaptive water management in societies and sustainability challenges (Table 1.1).

Agrarian Societies: Hydraulic Labour and Communal Ethics

In early riverine civilizations like Mesopotamia and the Indus Valley, water management necessitated collective labour, embedding values of cooperation and hierarchical governance. Large-scale irrigation fostered centralized authority, but it also cultivated a **moral economy of reciprocity**, as seen in Andean *ayni* (reciprocal labor for water infrastructure) (Trawick, 2003). Ancient civilizations pioneered sophisticated water technologies that shaped both their survival strategies and societal complexity. Shared water practices (Nile flood festivals) solidified social cohesion, transforming hydrological labor into sacred obligation (Némedi, 1995). Water mills, drainage systems, canals, and aqueducts—such as Rome's Aqua Appia (312 BCE) and China's Dujiangyan irrigation network (256 BCE), still operational today—demonstrated early mastery of hydraulic engineering and became the cultural heritage at the same time. Qanats, subterranean channels in Persia, tapped groundwater sustainably, while Mesopotamian levees and Egyptian nilometers optimized flood management. Yet these innovations carried hidden costs, for example the Roman use of lead pipes, for instance, likely contributed to widespread lead poisoning. Similarly, Mesopotamia's intensive irrigation caused soil salinization, collapsing agricultural productivity by 1700 BCE—an early lesson in ecological limits. Grand projects like the Ma'rib Dam in Yemen (eighth century BCE) faltered under silt accumulation and climate shifts, foreshadowing modern infrastructural vulnerabilities. Aesthetic displays, such as Hellenistic fountain houses, merged utility with cultural prestige, at the same time built on elite control over water access. These ancient systems reveal a paradox: technological brilliance often coexisted with environmental myopia, a theme echoing in today's debates over dam sustainability and resource equity.

Middle Centuries, Renaissance

The medieval Renaissance catalyzed advancements in hydraulic engineering, with innovations like improved aqueducts, watermills, and urban drainage systems. Cities such as Florence and Milan employed iron-reinforced infrastructure, exemplified by the *Navigli* canal network, which facilitated trade and urbanization (Gies & Gies, 1994). However, these systems often exacerbated social stratification: elite neighborhoods accessed clean water via private conduits, while poorer districts relied on polluted public wells, perpetuating health disparities. The Industrial Revolution intensified these dynamics. Centralized waterworks, like London's Bazalgette Sewers (1865), reduced cholera outbreaks but prioritized industrial zones, embedding class divides into urban planning. Meanwhile, hydropower dams, such as the Niagara Falls project (1895), powered factories but displaced Indigenous communities, illustrating **hydraulic colonialism** (Swyngedouw, 2015).

The shift to hydropower and fossil fuels in eighteenth-century Europe redefined values around productivity and ownership. **Protestant ethic** (Weber, 1905) emphasizes hard work, frugality, and discipline, which historically facilitated the accumulation of capital and thus contributed to the development and exploitation of energy resources within capitalist systems. Its linked Calvinist discipline to capitalist energy exploitation, as waterwheels and coal engines prioritized efficiency over sustainability. The enclosure of British commons for hydropower mills exemplified **Marx's primitive accumulation**, divorcing water from communal ethics and recasting it as a capitalist resource.

Modernity: Risk, Rights, and Environmental Justice

The twentieth century's mega-dams and fossil fuel crises spurred ethical reckonings and moving to recognize environmental priority with risks calculation. The concept of the **risk society** (Gabler, 2010) frames climate-driven water scarcity as a byproduct of industries, where marginalized groups—from Bangladesh's flood-displaced to Flint's lead-poisoned residents—bear disproportionate harm. The idea of **Elinor Ostrom's commons governance** (1990) revived communal ethics, inspiring Bolivia's *Law of Mother Earth* (2010), which grants legal rights to waterways (Boelens et al., 2016). The shift from waterwheels to hydroelectric dams mirrors **Kondratiev waves** of technological cycles. The nineteenth-century hydropower boom, epitomized by Niagara Falls' turbines (1895), catalyzed the **Second Industrial Revolution** (Smil, 2017), enabling urbanization. Vaclav Smil's **energy transitions** theory frames hydropower as a bridge from biomass to fossil fuels, yet its resurgence in the twenty-first century underscores unresolved tensions between renewable energy and ecosystem disruption.

The twentieth century saw hydrology emerge as a scientific discipline, formalizing water management through models like the Penman–Monteith equation (1965) for evapotranspiration. Yet, technocratic approaches often ignored local knowledge. Mega-dams like Egypt's Aswan High Dam (1970) boosted irrigation but triggered soil salinity and displaced 100,000 Nubians, fracturing social networks. Post-industrial cities, such as Los Angeles, epitomized the paradox of integrated systems: while the Los Angeles Aqueduct (1913) enabled sprawl, its dependence on distant water sources bred conflicts with Owens Valley communities, reshaping regional power dynamics (Reisner, 1986).

Contemporary challenges—polluted groundwater in Delhi (70% of supply contaminated) and Jakarta's subsidence crisis (25 cm/year)—highlight the limits of engineered solutions (Mukherjee et al., 2024). Conversely, decentralized initiatives like Barcelona's *super-illa* (superblocks) integrate green infrastructure, fostering community collaboration and redefining urban hydrosocial relations. These shifts underscore water's dual role as a driver of urbanization and a mirror of societal values—where infrastructure encodes power, equity, and ecological ethics drawing on Ian Morris's **energy capture approach** (Morris, 2010), which posits that a society's capacity to harness energy fundamentally structures its cultural priorities.

Future Paradigms: Sustainability and Planetary Ethics

As renewable energy transitions accelerate, values are pivoting toward **energy justice** (Sovacool, 2021) and **hydro-solidarity** (Sultana, 2022) with equite access of water and energy resources for humanity. Floating solar farms and green hydrogen projects reflect a **post-growth ethos**, prioritizing ecological limits over extraction. The UN's *Water Action Decade* (2018–2028) embodies **cosmopolitan ethics**, framing water as a global commons requiring transnational stewardship (Gareau & Crow, 2006). The concept of **planetary boundaries** (Rockström et al., 2009) highlighted positions water and energy systems within Earth's finite carrying capacity, demanding intergenerational equity. In this research was highlighted, that crossed the line of the critical planetary boundaries related to water is the disruption of the global hydrological cycle, which is exacerbated by climate change, leading to altered precipitation patterns, increased frequency of droughts, and more intense flooding events. Freshwater use and pollution are pressing concerns, as excessive consumption and contamination of water resources threaten biodiversity, ecosystem health, and human access to clean water. Over-extraction of groundwater for agricultural, industrial, and domestic purposes is depleting aquifers faster than they can be replenished, posing a significant risk to water security and sustainable development.

The metabolic rift illuminates how industrial capitalism severed human-water symbiosis, commodifying rivers for energy extraction. The transition from biomass to hydropower and fossil fuels reconfigured labour relations and class structures. The **metabolic rift** theory (Foster, 1999) argues that industrial capitalism severed humans from ecological cycles, commodifying water and energy to fuel urbanization. In nineteenth-century Manchester, hydropowered mills concentrated workers in squalid factory towns. Today, **energy justice** scholars (Sovacool, 2021) highlight how mega-dams—like Brazil's Belo Monte—displace Indigenous communities while subsidizing energy for urban industries, perpetuating **neo-colonial extractivism**. Colonial projects, such as British dam-building in India, exemplify **hydro-hegemony**

(Zeitoun & Warner, 2006), where water control reinforces geopolitical dominance. Conversely, Elinor Ostrom's **commons approach** (Herzberg, 2020) challenges this, showcasing communal irrigation systems (*acequias* in Spain, *subak* in Bali) that thrived without state coercion. Modern dam-building can reflects terraforming is the process of modifying the atmosphere, temperature, surface topography, or ecology of a planet, moon, or other celestial body to make it habitable for Earth-like life (WOODS, 2019), where human agency reshapes planetary systems, often exacerbating ecological fragility (for example, Aral Sea desiccation).

The management of water and energy has never been a neutral technical endeavor; it is a sociopolitical process that crystallizes power relations, molds institutions, and perpetuates inequalities. From the earliest agrarian societies to modern megacities, control over these resources has functioned as both a catalyst for collective action and a mechanism for stratification. To unravel this duality, contemporary sociological theories—from **structuration** to **political ecology**—offer critical lenses to decode how hydrological and energy systems shape social hierarchies, institutional practices, and cultural norms.

The **hydraulic hypothesis** (Wittfogel, 1957) poited that large-scale irrigation systems necessitated centralized bureaucracies, creating "hydraulic civilizations" finds resonance in the **resource curse** framework (Ross, 1999), where concentrated control over water or energy fosters rent-seeking elites. For example, the **Nile's annual inundation** was not merely an ecological event but a political ritual: Pharaohs claimed divine authority by "regulating" the flood, embedding hydraulic management into Egypt's theocratic statecraft (Mikhail, 2017) Modern parallels emerge in urban water apartheid, as seen in Cape Town's 2018 "Day Zero" crisis, where privatized desalination projects prioritized wealthy enclaves while informal settlements faced rationing.

Water and energy systems institutionalize specific social practices. **Structuration approach** (Whittington, 2015) explains how dam-building practices —from public consultations to environmental impact assessments—reproduce bureaucratic norms, legitimizing state authority. Conversely, **Ostrom's polycentric governance** (1990) underscores grassroots systems like Bolivia's *regantes* (water user associations), where communal irrigation rules foster collective identity and reduce elite capture.

In the twenty-first century, smart water grids and decentralized renewables dissolve boundaries between nature and technology, creating **hydro-social systems** (Swyngedouw, 2015) (Singapore's **PUB Smart Water Grid** uses AI to optimize consumption, embedding neoliberal rationality into everyday practices. Yet, such techno-utopias risk **algorithmic bias**: predictive models may prioritize affluent districts, exacerbating inequalities. Climate-driven water scarcity as a manufactured risk, where elite "adaptation enclaves" (e.g., Dubai's cloud-seeding projects) privatize resilience, leaving marginalized populations exposed. Meanwhile, **platform hydrocapitalism** (*Hydroimperialism, Hydrocapitalism, Communism and Crises*, n.d.) commodifies data: companies like Xylem Inc. sell IoT-based water analytics, transforming public goods into profit streams.

Yet, the persistence of "hydro-hegemony" in transboundary disputes and the commodification of water data through platform hydrocapitalism reveal entrenched power asymmetries. Hydrosocial constructs shaped by colonial legacies, metabolic rifts, and evolving notions of justice. To align with Rockström's planetary boundaries and Foster's metabolic restoration, future energy systems must transcend techno-optimism, prioritizing decommodification, participatory design, and reparative justice for communities displaced by decades of extractive infrastructure. This requires reimagining energy sovereignty as inseparable from ecological stewardship—a synthesis of historical reckoning and radical innovation that redefines sustainability as a practice of collective liberation rather than mere resource efficiency.

In this book the authors try to go out through disciplinary boundaries. The "Chapter 2. Modern theoretical approaches and methodology for water and hydro energy systems "will provide a comprehensive analysis of contemporary approaches to studying water resources and hydropower systems, focusing on their technological, economic, and sociopolitical dimensions. The technological potential of hydropower, wave energy, and tidal energy will be examined alongside innovations in energy extraction, storage, and distribution in "Chapter 4. Sustainable Marine and Offshore Energy Management" and energy security in

Chapter 3. "Demand Management Instruments for Water and Energy Security in the SDG Era". Economic instruments for financing sustainable infrastructure and market-driven solutions will be evaluated in "Chapter 5: Circular Water Economy and the Water-Energy Nexus", as will the role of social energy policies in addressing equity and accessibility in case "Chapter 7. "Exploring the complex interdependence nature of marine renewable energy sector: a developing country perspective". Additionally, the chapter will explore energy-saving strategies across industries in "Chapter 9. Industrial and Agricultural Water Conservation and Energy", emphasizing the integration of water-energy systems into broader climate resilience frameworks and sustainable behavior in "Chapter 6. Water-Energy Consumption and Sustainable Behavior Practices". The cities landscape will be studied in "Chapter 8. Renewable Energy Innovations in Urban Sustainability: Water and Energy Synergies". In synthesizing these elements, the analysis aims to highlight interdisciplinary pathways for balancing ecological sustainability, economic viability, and social justice in the evolving global hydro energy landscape.

References

Boelens, R., Hoogesteger, J., Swyngedouw, E., Vos, J., & Wester, P. (2016). Hydrosocial territories: A political ecology perspective. *Water International, 41*, 1–14. https://doi.org/10.1080/02508060.2016.1134898

Foster, J. B. (1999). Marx's theory of metabolic rift: Classical foundations for environmental sociology. *American Journal of Sociology, 105*(2), 366–405. JSTOR. https://doi.org/10.1086/210315

Gabler, M. (2010). World at risk, edited by Ulrich Beck. English Edition, Translated by Ciaran Cronin (9. 269). Polity Press, 2009, $24.95, Paperback. *European Journal of Risk Regulation, 1*(2), 201–203. Cambridge Core. https://doi.org/10.1017/S1867299X00000362

Gareau, B., & Crow, B. (2006). Ken conca, governing water: Contentious transnational politics and global institution building. *International Environmental Agreements.* https://doi.org/10.1007/s10784-006-9007-1

Gies, F., & Gies, J. (1994). *Cathedral, forge, and waterwheel: Technology and invention in the middle ages*. HarperCollins.

Herzberg, R. Q. (2020). Elinor Ostrom's governing the commons. *The Independent Review, 24*(4), 627–636. JSTOR.

Hydroimperialism, Hydrocapitalism, Communism and Crises. (n.d.). CIUHCT. Retrieved April 7, 2025, from https://ciuhct.org/en/meetings/hydroimperialism-hydrocapitalism-communism-and-crises-1

Isendahl, C., & Scarborough, V. (2004). The flow of power: Ancient water systems and landscapes. *Latin American Antiquity, 15*, 117. https://doi.org/10.2307/4141571

Mikhail, A. (2017). *Under Osman's tree: The Ottoman Empire, Egypt, and environmental history*. University of Chicago Press.

Morris, I. (2010). *Why the west rules—for now: The patterns of history, and what they reveal about the future*. Farrar.

Mukherjee, P., Sunar, S., Saha, P., Saha, S., Dutta, S., & Yakub Ali, S. (2024). An integrated approach towards groundwater quality and human health risk assessment in the Indo-Gangetic plains of West Bengal, India. *Environmental Nanotechnology, Monitoring & Management, 22*, Article 101022. https://doi.org/10.1016/j.enmm.2024.101022

Némedi, D. (1995). Collective consciousness, morphology, and collective representations: Durkheim's sociology of knowledge, 1894–1900. *Sociological Perspectives, 38*(1), 41–56. JSTOR. https://doi.org/10.2307/1389261

Reisner, M. (1986). *Cadillac desert: The American West and its disappearing water*. Viking.

Rockström, J., Steffen, W., Noone, K., Persson, Å., Chapin, F. S., Lambin, E. F., Lenton, T. M., Scheffer, M., Folke, C., Schellnhuber, H. J., Nykvist, B., de Wit, C. A., Hughes, T., van der Leeuw, S., Rodhe, H., Sörlin, S., Snyder, P. K., Costanza, R., Svedin, U., & Foley, J. A. (2009). A safe operating space for humanity. *Nature, 461*(7263), 472–475. https://doi.org/10.1038/461472a

Ross, M. (1999). The political economy of the resource curse. *World Politics, 51*, 297–322. https://doi.org/10.1017/S0043887100008200

Smil, V. (2017). *Energy and civilization: A history*. MIT Press.

Sovacool, B. K. (2021). Who are the victims of low-carbon transitions? Towards a political ecology of climate change mitigation. *Energy Research & Social Science, 73*, Article 101916. https://doi.org/10.1016/j.erss.2021.101916

Sultana, F. (2022). The unbearable heaviness of climate coloniality. *Political Geography, 99*, Article 102638. https://doi.org/10.1016/j.polgeo.2022.102638

Swyngedouw, E. (2015). *Liquid power: Contested hydro-modernities in twentieth-century Spain.* MIT Press.

Trawick, P. (2003). Against the privatization of water: An indigenous model for improving existing laws and successfully governing the commons. *World Development, 31*(6), 977–996. https://doi.org/10.1016/S0305-750 X(03)00049-4

Weber, M. (1905). *The protestant ethic and the spirit of capitalism.* Routledge.

Whittington, R. (2015). Giddens, structuration theory and strategy as practice. In D. Golsorkhi, L. Rouleau, D. Seidl, & E. Vaara (Eds.), *Cambridge handbook of strategy as practice* (2nd ed., pp. 145–164). Cambridge University Press; Cambridge Core. https://doi.org/10.1017/CBO9781139681032.009

Wittfogel, K. A. (1957). *Oriental despotism: A comparative study of total power.* Yale University Press. Retrieved from https://www.amazon.com/Oriental-Despotism-Comparative-Study-Total/dp/0300010540

WOODS, D. (2019). Terraforming earth. *Diacritics, 47*(3), 6–29. JSTOR.

Zeitoun, M., & Warner, J. (2006). Hydro-hegemony—a framework for analysis of trans-boundary water conflicts. *Water Policy, 8*, 5. https://doi.org/10.2166/wp.2006.054

2

Modern Theoretical Approaches and Methodology for Water and Hydro Energy Systems

Pius Siakwah and Obodai Torto

Abstract Over the years, sectoral analysis has overlooked the complex relationships between water and energy systems in development. Allowing silo policies silos to thrive. Yet, the complexities of development require an appreciation of the theoretical and methodological link between water and energy systems and policy implications. This chapter explores the contemporary theoretical frameworks and methodologies for analysing water and energy systems to provide a comprehensive understanding of the multifaceted relationships for sustainable development. It integrates insights from the environmental, economic, and social sciences—transdisciplinary approaches, such as nexus, multi-level perspective, system-based analysis, flow analysis, and the Sustainable Development Goals (SDGs). Transdisciplinary approaches enable diverse actors and institutions to engage in collaborative innovations beyond discipline-specific boundaries in water and energy policy. It provides a holistic perspective for knowledge and skills sharing and applications to achieve water and energy policy goals while bringing to the fore the competing interests in such complex systems.

© The Author(s), under exclusive license to Springer Nature Switzerland AG 2025
Y. Ermolaeva et al., *Rethinking Water and Energy for a Sustainable Future*, Palgrave Studies in Climate Resilient Societies, https://doi.org/10.1007/978-3-032-04485-3_2

Keywords Water and energy policy · Interdisciplinarity · Nexus · System-thinking · Flow analysis · And sustainability

Introduction: A Transdisciplinary Approach to Water and Energy Policy

Natural resources and their uses have engaged the attention of development, research and policy experts. For example, water and energy resources have been at the fore of some of Africa's sustainable development discourses (Elmqvist et al., 2014; Farmandeh et al., 2024; Kumar & Karmakar, 2024). The complex relations among these resources have also raised concerns about transdisciplinary methodologies and theories such as nexus, flow and system analysis. Some of these nexuses emerged globally to mitigate the exigencies of climate change and social changes, including population growth and urbanization. For example, by the year 2050, over 50% of the global population will be urbanized, and together with population and economic growth, will put significant pressure on water, energy, food, and land resources (Dai et al., 2018). Systems analysis shows how pressure on water, energy, and food resources emanates from multiple sources. The dynamics of water, energy, and food interlinkages can impede sustainable development (Ali & Acquaye, 2024, 2). Increasing demand for alternative energy, industrial reforms and climate change discourses complicate water-energy policies (Hamiche et al., 2016). Understanding the complex water and energy methodological and theoretical perspective requires looking at how global, national, and local policies and practices shape development outcomes.

Sustainable Development Goals (SDGs) and the Paris Agreement are global initiatives that continue to shape water and energy policy. The 2015 Paris Climate Agreement aim to curb carbon dioxide emissions by 5 billion tonnes by 2030 and net-zero by 2050 (see Ali & Acquaye, 2024). National energy systems change support the achievement of net-zero emissions (Rogelj et al., 2015). The SDGs are an improvement of the Millennium Development Goals (MDGs) (Paoli & Addeo, 2019). The SDGs highlight how various development goals are related. The complex systems approach helps in analysing the relations among multi-layered sectors. Changes include the advancement of renewable energy sources, integration of energy systems, and social system changes relative to attitudes and values.

Transdisciplinarity helps to appreciate the complex relations between water and energy systems and policy initiatives by highlighting the trade-offs and synergies (Mei et al., 2021; Endo et al., 2020). It is where researchers from different fields collaborate with diverse theoretical and methodological strategies and perspectives to address complex problems through innovations (Botai et al., 2021; Foran, 2015). This is also related to cooperation with diverse groups and stakeholders across sectors, translating systems thinking into policymaking processes and balancing different user goals and interests. This approach emphasizes a holistic perspective in knowledge and skills sharing and applications. It is a comprehensive approach to understanding the multifaceted sustainable development systems (Schwanen, 2018).

The concept is transdisciplinarity has manifested significantly in the water-energy-food (WEF) nexus and policy implications. (Ali & Acquaye, 2024; Dai et al., 2018). Allouche et al. (2019) explore the knowledge nexus and transdisciplinarity, concentrating on interconnections among and between nature, society and technology (Schwanen, 2018). Transdisciplinarity reflects an admonition for the integration of research efforts and policy prescriptions across disciplines (Endo et al., 2017; Stirling, 2015). In Africa, transdisciplinary approaches have evolved (Botai et al., 2021; Endo et al., 2020) to shape water and energy policies, highlighting power relations, historical, cultural and socio-political dimensions of the relationships (Foran, 2015) to account for winners and losers.

The broader question is how modern scientific transdisciplinary or interdisciplinary approaches enhance a more nuanced understanding of water and energy systems. What are the challenges and opportunities associated with incorporating transdisciplinary theoretical approaches and methodologies in analysing water and energy systems? This chapter examines how modern transdisciplinary approaches enhance the understanding of water and energy systems by integrating environmental, economic, and social sciences. We focus on nexus, multi-level perspective

(MLP), system-based analysis, flow analysis, and SDGs and sustainability as transdisciplinary approaches. The transdisciplinary tools help to mitigate the complex challenges in water and energy management, resource efficiency, and environmental conservation. The chapter proceeds by discussing nexus approaches, multi-level perspectives, flow analysis, system-based analysis and SGDs and sustainability.

Nexus and Water and Energy Policy—Theory and Methodology

The nexus concept has been applied to various thematic areas to understand the complex human relations, problems and phenomena (Estoque, 2023; Harwood, 2018; Hoff, 2011; Pahl-Wostl, 2019; Rasul & Sharma, 2016; Weitz et al., 2014). Some of these nexuses include: development-security nexus, development-migration, water-energy, water-food, energy-food, and more complex ones such as water-energy-food (WEF). The application of nexus, be it methodological or theoretical, is to appreciate the complex relations between phenomena, identify opportunity costs and promote sustainability.

Since 2011, transdisciplinarity research has proliferated, reflecting a call for integration of research efforts and policy prescriptions across disciplines and sectors (Endo et al., 2017; Stirling, 2015). The WEF nexus concept has a strong security focus, promoted at the World Economic Forum in 2008 held in Davos, Switzerland (Pahl-Wostl, 2019; Waughray, 2011). The Germans promoted WEF in the run-up to the Rio+20 sustainability summit, '*The Water-Energy-Food Security Nexus: Solutions for the Green Economy*' in Bonn, 2011 (Hoff, 2011; Pahl-Wostl, 2019). The Stockholm Environment Institute also promoted WEF prior to the Bonn Conference in 2011 (Ali & Acquaye, 2024).

Nexus represents multi-dimensional scientific inquiry into the complex and non-linear interactions between water, energy, and food and societal implications (Allouche et al., 2019). The WEF nexus approach helps to overcome governance failures in resource management (Farmandeh et al., 2024; Pahl-Wostl, 2019). We illustrate the nexus as a complex schema (see Fig. 2.1) of the water-energy-food system (Shannak et al., 2018).

The schemata incorporate political, social and economic factors to address the complex multi-sectoral issues (Shannak et al., 2018). Nexus thinking involves recognizing and dealing with the interactions and interconnections between multiple sectors, including related synergies, conflicts, and trade-offs (Darko et al., 2023; De Loë & Patterson, 2017; Jackson, 2024; Simpson & Jewitt, 2019). The relations are sometimes contradictory, complementary and collaborative.

The nexus can serve as a theoretical (conceptual) or methodological tool. As a concept, nexus is an integrated framework that guides policy-makers in analysing and executing sustainable actions (Ali & Acquaye, 2024). The synergy in nexus promotes a move from silo governance to holistic and complex governance approaches to sustainability (Darko et al., 2023). This is related to integrative frameworks that support the creation of compatible datasets to support decision-making, through interactive platforms and maps (Allouche et al., 2019). De Strasser et al. (2016) explore a transboundary river basin nexus approach by practically assessing selected basins to identify trade-offs and impacts across sectors and countries.

Fernandes Torres et al. (2019) propose a multisectoral systematic literature review procedure to develop 'nexus thinking' through (a) understanding nexus thinking, (b) identification of composing variables, (c) evaluation (diagnosis and prognosis), and (d) decision-making. The quantitative methods help to identify variables, problem evaluation and system simulation, while qualitative methods promote description (Endo et al., 2015). Farmandeh et al. (2024) integrate content analysis with analytical network process to investigate the what, why, and how of nexus.

There are opportunities and challenges associated with the nexus. WEF develops socially and politically relevant policies that transcend the sectoral domains (Albrecht et al., 2018). It challenges global governance and policies that have often focused on a silo governance regime (Ali & Acquaye, 2024) while enhancing efficiency and cost-effectiveness

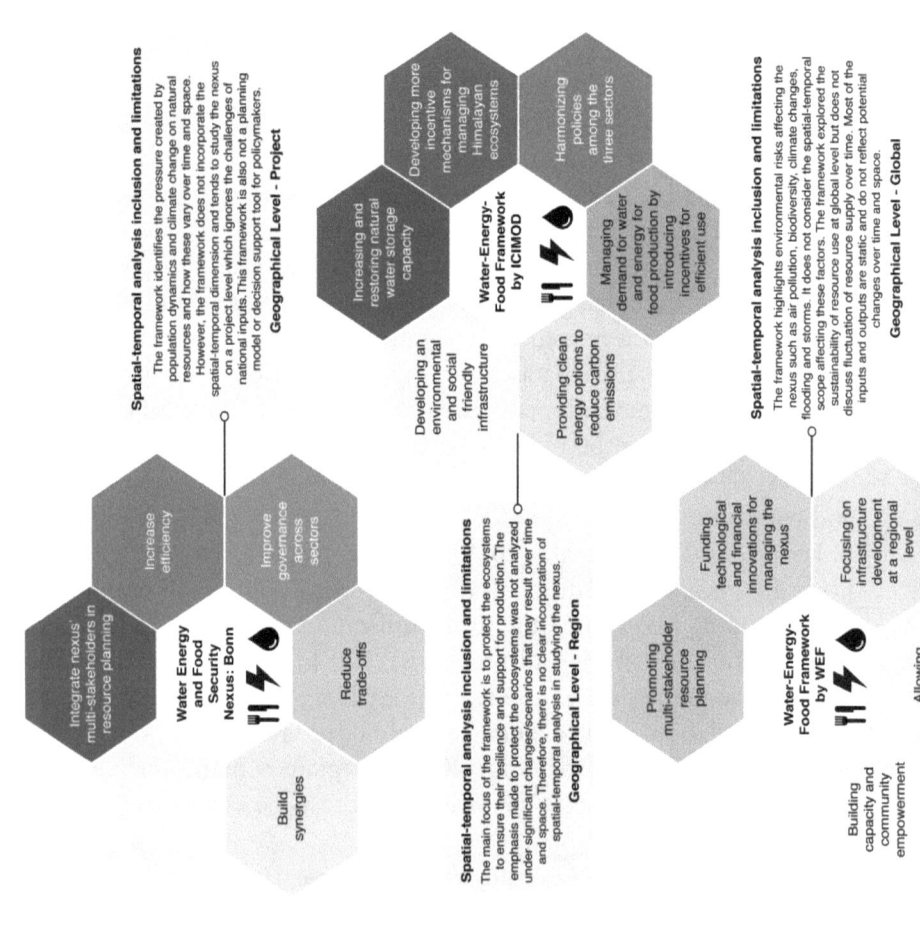

in planning and implementation (Leck et al., 2015). It addresses fragmentation of policies and institutions across the water, energy, and food sectors (Darko et al., 2023).

The WEF nexus is overly focused on how things ought to be rather than how they are. Thus, it explains ideal systems, whether energy or water or the links among them. Shannak et al. (2018) rightly noted the need to move from theory to practice of the WEF by evaluating existing frameworks. Additionally, nexus is rooted in the landscape with a limited sympathy for local dynamics (De Strasser et al., 2016). Often, even where nexus is holistically, one sector still dominates (Darko et al., 2023; Leck et al., 2015), mostly the first theme. Some of the nexus approaches are quantitatively biased, with over 70% focusing on input–output analysis, trade-off analysis, and scenario analysis (Albrecht et al., 2018). WEF approaches are poorly politicized or historicized.

Multi-level Perspective and Energy-Water Policy

The multi-level perspective (MLP), as part of the broader socio-technical transitions research, emerged in the early 2000s in the field of innovation studies (Geels, 2019). MLP has roots in the works of Rip and Kemp (1998), refined by others (Geels, 2005; Geels & Schot, 2007). It is associated with urban services to interrogate large-scale and centralized infrastructure regimes (Markard et al., 2012). MLP conceptualizes overall dynamic patterns in socio-technical transitions (Geels, 2004, 2011, 2019; Merton, 1968).

MLP conceptualizes energy and water systems by assessing complex interactions among institutions and actors across three scales: landscape, regime, and niche (Geels, 2002, 2005, 2019) as illustrated in Fig. 2.2 below. This is a multidimensional transition framework (Geels, 2018) to understand how socio-technical transitions occur, focusing on the interplay between niche innovations, socio-technical regimes, and socio-technical landscapes to understand transformational change.

Landscape encompasses global-level ideologies, institutions, discourses, political ideologies, societal values, and economic trends

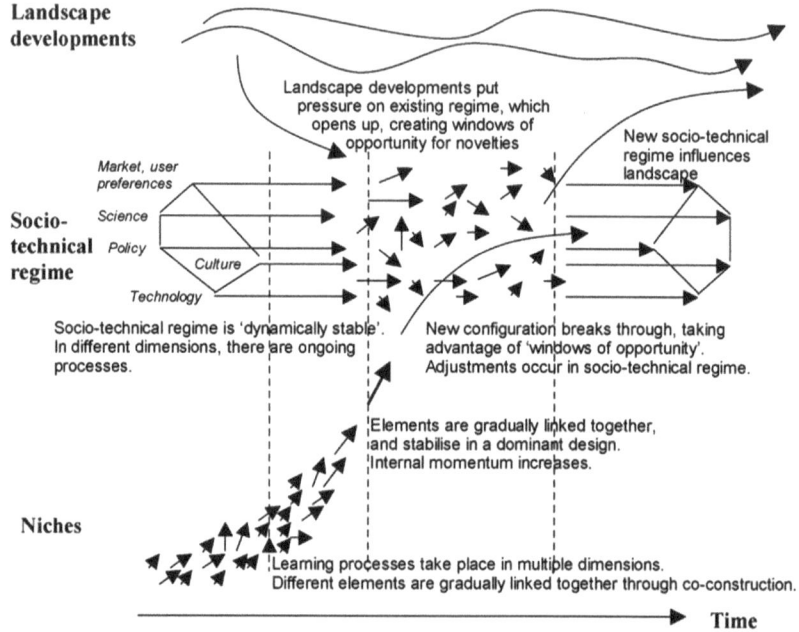

Fig. 2.2 Multi-Level Perspective (MLP)—socio-technical transitions. *Source* Geels (2002, 2019)

at the macro level (Geels, 2011; Power et al., 2016), which impact the dynamics of regimes and niches. These are the entrenched structural societal relationships that change only gradually.

The socio-technical regime forms the second layer of "deep structures" that accounts for the stability of an existing socio-technical system (Geels, 2004). It is the semi-coherent set of rules that orient and coordinate the activities of the social groups that reproduce the various elements of socio-technical systems (Geels, 2011). Due to lock-ins in the regimes, innovation occurs incrementally.

Niches are "protected spaces" such as research and development laboratories, subsidized demonstration projects, or small market niches for emerging innovations (Geels, 2011). Niche actors (e.g., entrepreneurs and start-ups) work on radical innovations that deviate from existing regimes. Rip and Kemp (1998), Schot and Geels (2013) distinguish three core processes in niche development:

- Articulation of expectations or visions which guide the innovation activities to attract attention and funding from external actors.
- Building of social networks and the enrolment of actors to expand the resource base for innovations.
- Learning and articulation processes relative to technical design, market, infrastructure, business models, and policy.

Niches are protected spaces for innovation (Smith & Raven, 2012), while the regime constitutes a set of rules, norms and institutions, and landscapes are the exogenous political developments, social relationships which exert pressure on the regimes (Oates, 2021). MLP emphasizes radical innovations, enacted by multiple social groups (Geels, 2019), within the context of nested hierarchy. Thus, transitions are non-linear processes resulting from an interplay at three analytical levels (Geels, 2002; Rip & Kemp, 1998).

Geels (2019, 187) recently re-conceptualized the MLP within the context of politics and power, cultural discourse and struggles, grassroots innovation, multiple transition pathways, resistance and reorientation, destabilization and decline, and policy analysis. These are important given evolving sustainability discourses relative to energy and water.

Bakhuis et al. (2024) identify 'multi-system frameworks', similar to the 'multi-level governance' (Tasan-Kok & Vranken, 2011), as variants of the multi-level perspective. This entails making water and energy decisions that engage a variety of politically independent but interdependent actors across diverse levels. There are non-state actors in the energy transition discourse, mostly driving niche innovation (Oates, 2021) in the Global South. Francis et al. (2022) explore issues of 'network governance' in the renewable energy transition in Africa, focusing on collaborative strategies for planning and decision-making.

The MLP concept has come with criticisms for a lack of agency, largely explanatory, and flat ontologies despite the hierarchical nature (Geels, 2011). Smith et al. (2005) view MLP as descriptive and with limited room for agency.

System-Based Analysis of Water and Energy Policy

System-based analysis is both a theoretical and methodological approach that examines the complex relationships among parts and how these parts interact in the wider system. It involves breaking down a system into its components, goals and optimizing performance. System-based analysis operates similarly to multi-level systems (Geels, 2004, 2011), but it is not hierarchical. Water and energy policies can be analysed as a system, looking at how the parts of the water and energy systems interact.

Energy system models are characterized by multiple uncertainties in production, distribution and trade-offs in transition scenarios (Pye et al., 2015). The challenge of, however, understanding, assessing and communicating uncertainties is magnified by model sophistication (Davies et al., 2014). The energy systems models provide pathways toward low-carbon (Chang et al., 2021; Del Granado et al., 2018). Chang et al. (2021) observe that modeling the energy transition system is crucial to planning. Energy transition and economic models help operate as top-down and bottom-up models in terms of generation and demand, grid operations, infrastructure and macroeconomic interactions (Del Granado et al., 2018). Water and energy system-based analysis helps to understand how dams serve as energy generation and a water source.

Davidsdottir et al. (2024) stakeholder engagement is important to capture the complexity of the net-zero transition. Such analysis focuses on the complex relations between social impact, economic development, environmental impact, energy security, and technical aspects (Davidsdottir et al., 2024). The integrated energy system (IES), which consists of renewable energy generation systems, heating and power systems, has become important in global energy policy (Correa-Posada & Sanchez-Martin, 2014; Ma et al., 2017). A system approach to water-energy improves efficiency and sustainability.

Methodological approach to system analysis integrates technical descriptions and application aspects of the modeling tools, such as policy-relevance, user accessibility, and model linkages (Chang et al., 2021). Kusakana (2014) explores the techno-economic analysis of off-grid hydrokinetic-based hybrid energy systems for rural South Africa. Such a system provides a low-cost and sustainable electrical energy supply to isolated rural areas of South Africa with adequate water resources. This system-based analysis improves efficiency, optimizing processes and resource allocation, enhances decision-making, reduces costs, eliminates inefficiencies, and gains a deeper understanding of complex systems and opportunities for innovation.

Flow Analysis and Water-Energy Policy

Flow analysis is one of the transdisciplinary techniques for understanding water-energy and policy, mostly through quantitative analysis to interrogate the relations among units in energy systems from production, distribution, use and ecological implications. It encompasses the movement of resources, information, or data through a system or process (Chen & Chen, 2015; Chen et al., 2016; Li et al., 2022). Flow analysis is not new (Cerdà et al., 1999) is a system analysis.

Recent flow analysis techniques, including energy flow analysis (EFA), input-out analysis (IOA), and ecological network analysis (ENA), help in tracking the energy balance across countries, cities and industrial units (Chen & Chen, 2014, 2015; Swilling, 2006). These are more comprehensive and balanced techniques for understanding urban energy consumption through integrating accounting perspectives into the flow of resources across systems. The EFA quantifies both the primary and final energy consumption of the urban economy, while ENA quantifies ecological issues relative to city energy activities (Chen & Chen, 2015, 99). The ENA approach understands the system as an integration of smaller interconnected entities (Chen & Chen, 2015).

Flow analysis can be used to understand water and energy use in various sectors through understanding energy production, transmission, consumption and ecological implications. To understand an energy flow profile, the urban economy is modeled as the aggregation of a set of interconnected economic sectors (Chen & Chen, 2015). This helps to track the energy balance of nations, cities and industrial units (Chen & Chen, 2014; Swilling, 2006).

Material (energy) flow analysis (MFA) has evolved from its early applications in the 1970s to become a crucial tool in energy policy and sustainable development (Chen & Chen, 2015; Kullmann et al., 2021). While it initially focused on quantifying material flows relative to economic activities (Chen & Chen, 2015), it has expanded to include energy flows and environmental and, highlighting interconnectedness between society, nature, and the economy.

We can view the flow analysis in the context of input–output analysis (see Fig. 2.3). Chen and Chen (2015) analyse urban energy consumption based on input–output analysis. It accounts for energy consumption (direct and indirect) used in the production of goods and services.

Accounting for energy consumption is important for sustainable energy planning of cities for several reasons (Chen & Chen, 2015; Sampaio et al., 2013). It provides insight into the level of energy consumption in different urban activities and future performance based on demand and technology, consumption and relationships between urban sectors (Chen & Chen, 2015). Additionally, it provides a mechanism for selecting the optimal energy systems based on environmental and economic conditions. Appreciating the complex relations between water and energy helps to understand decarbonizing power generation, compliance with environmental needs, optimizing infrastructure use and diversification of power generation. For example, Sterl et al. (2021) show how the multiple political and environmental challenges surrounding GERD (Grand Ethiopian Renaissance Dam) are mitigated by balancing the needs and aspirations of Ethiopia and Egypt.

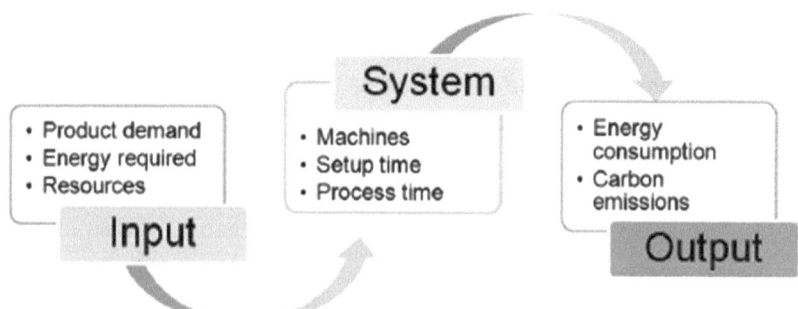

Fig. 2.3 Input–output analysis. Anagnostakis (2015)

SDGs as a Conceptual Approach to Water-Energy Policy

Sustainability can be used as both a conceptual tool for analysing development while serving as a methodological mechanism for assessment. The concept of sustainable development has evolved over the years. Indeed, every society, even if not overt, in its activities, has included practices that seek to meet present and future needs. However, due to increasing population, consumption and industrial activities, the issue of sustainability, especially relative to climate change, has become topical (see Eberling & Langkau, 2024; Emina, 2021; Jain & Jain, 2020; Elmqvist et al., 2014; Meadowcroft, 2007). The Brundtland Report in 1987, partly birthed the sustainable development discourses on the global stage, and conceptualizes it as development that meets the needs of current generations without compromising the ability of future generations to meet their own needs (WCED, 1987). Sustainable development core principles include long-term aspiration of striking a just balance between the economy, environment and society relative to the present and future. Sustainable development seeks human well-being without stretching the ecological limits (Jain & Jain, 2020). Within the frame of water-energy policy, we can extend it to the sustainable use of water resources—blue economies (Sungkawati, 2024) and techno-economic systems (Kumar & Karmakar, 2024).

The Sustainable Development Goals (SDGs), also known as the Global Goals, were crystallized and adopted by all United Nations member states in 2015 to sustainably improve human and environmental well-being. The SDGs are a set of 17 interconnected goals as part of the 2030 Agenda for Sustainable Development aim to end poverty, protect the planet, and ensure all people enjoy health, justice, and prosperity by 2030—"leaving no one behind". It is the post-2015 Development Agenda 2030—a plan of action for people, the planet and prosperity (Elmqvist et al., 2014). The SDGs that specifically deal with sustainability relative to water and energy are #6 (clean water and sanitation), #7 (affordable and clean energy), #11 (sustainable cities and communities), #12 (responsible consumption and production), and # 13 (climate action) (Barbier et al., 2017).

Within the context of a transdisciplinary approach, the SDGs can be conceptualized as a systems approach to sustainability. Barbier and Burgess (2017) advocate for a systems approach to sustainability within the context of the SDGs. We have to look at the link between the systems approach to sustainability and the 17 Goals. The systems approach depicts sustainable development as the intersection of the goals: environmental (ecological), economic and social (Barbier & Burgess, 2017) and the trade-offs.

The *systems approach* characterizes sustainability as the maximization of goals across environmental, economic and social systems (Barbier, 2016; Costanza et al., 2016). The three systems in development require an adaptive process of trade-offs (Barbier & Burgess, 2017). Each of the SDGs is either an economic, environmental or social goal, but the systems approach allows us to analyse the UN SDGs as a systems approach to sustainability (Barbier & Burgess, 2017) as shown in Fig. 2.4.

Sustainability within the context of SDGs and development helps in understanding the multifaceted relationship between water and energy systems and other development indicators. Sustainability transitions entail interactions between technology, power, politics, economics, and public discourse. We, therefore, need theoretical approaches that address the multi-dimensional nature of sustainability transitions and dynamics of structural change (Geels, 2011). Also, institutional commitments, shared beliefs and discourses, power relations, and politics can stabilise existing systems (Unruh, 2000).

SDGs as a methodology strategy for assessing sustainability. Jain and Jain (2020) examine the policy perspectives on SDGs, asking to what extent do the 17 SDGs lead to enhancing human well-being [measured through Human Development Index (HDI)], considering the planet's carrying capacity [measured through Ecological Footprints (EF)] (Jain & Jain, 2020). Such an approach to SDG analysis suggests that countries which rank high on HDI are also high on the EF, and these countries have scored high on the SDG Index (Jain & Jain, 2020). Africa lags most significantly behind in its achievement of the SDGs, with a lower percentage of countries achieving the target by 2030 (Mahlatsi, 2021). Paoli and Addeo (2019) explore assessing SDGs as a methodology to

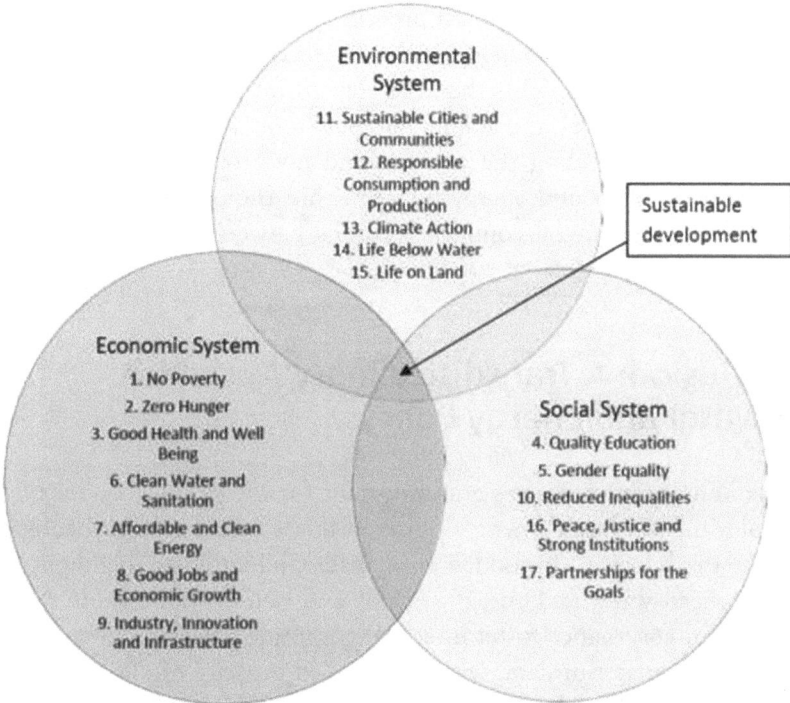

Fig. 2.4 Systems approach to SDGs. *Source* Barbier and Burgess (2017)

measure sustainability through the SDG composite index. This is based on cross-country national data and uses multivariate statistics to build composite indexes to assess country-level SDGs performance (Paoli & Addeo, 2019). SDG is a multi-pillar model for assessing sustainability (Paoli & Addeo, 2019). This originates from a growing concern about the three-pillar model (economic, environmental, and social) considered as overlooking other pillars of fundamental importance (Littig and Grießler, 2005; Dahl, 2012). It requires refining the operational definition of the SDGs to improve their reliability (Murphy, 2012).

Sustainability assessment methodologies are important tools for supporting decision-making on sustainable development measures (Sala et al., 2015). However, they comprise a broad variety of different methodologies, and no standard procedure has yet been defined (Eberling & Langkau, 2024). The diversity of approaches of the SDGs and the diversity of goals highlights the need for critical reflexivity and triangulation of data sources and methodological tools (Eberling & Langkau, 2024) in understanding the concept and how to achieve it. Thus, in interrogating water and energy policy within the context of transdisciplinary approaches, the complex relations between the SDGs provide useful insight.

Conclusion: A Transdisciplinary Approach to Water and Energy Policy

Water and energy resources are important for the global economy. The complexities of development require an appreciation of the theoretical and methodological, grounded in transdisciplinarity, to highlight the link between water and energy systems and policy implications. Nexus and system governance foster interlinkages, trade-offs and synergies. For example, Ghana's Bui dam, with an installed capacity of 400 MW, only produces 60 MW of hydropower (Siakwah & Torto, 2022). The irrigation component of the dam is not well developed. The water and energy policy around the Bui project in Ghana shows tensions, sidelining, and role conflicts among the actors, where the energy sector actors weigh power and resources over other institutions and actors (Siakwah & Torto, 2022).

This chapter argues that the contemporary theoretical frameworks and methodologies for analysing water and energy systems provide a comprehensive understanding of the multifaceted relationships for sustainable development. Integrating insights from the environmental, economic, and social sciences, in the context of transdisciplinary approaches, such as nexus, multi-level perspective, and system-based analysis, helps to interrogate the complex relations between water and energy systems. Transdisciplinary approaches enable diverse actors and institutions to collaborate in addressing complex problems through creative innovations beyond discipline-specific boundaries. It provides a holistic perspective for knowledge and skills sharing and applications to achieve water and energy policy goals while bringing to the fore the competing interests in such complex systems.

References

Albrecht, T. R., Crootof, A., & Scott, C. A. (2018). The water-energy-food nexus: A systematic review of methods for nexus assessment. *Environmental Research Letters, 13*(4), Article 043002.

Ali, S. M., & Acquaye, A. (2024). An examination of water-energy-food nexus: From theory to application. *Renewable and Sustainable Energy Reviews, 202*, Article 114669.

Allouche, J., Middleton, C., & Gyawali, D. (2019). *The water–food–energy nexus: Power, politics, and justice.* Routledge.

Anagnostakis, M. J. (2015). Environmental radioactivity measurements and applications–difficulties, current status and future trends. *Radiation Physics and Chemistry, 116*, 3-7.

Bakhuis, J., Kamp, L. M., Barbour, N., & Chappin, É. J. L. (2024). Frameworks for multi-system innovation analysis from a sociotechnical perspective: A systematic literature review. *Technological Forecasting and Social Change, 201*, Article 123266.

Barbier, E. B. (2016). Sustainability and development. *Annual Review of Resource Economics, 8*(1), 261–280.

Barbier, E. B., & Burgess, J. C. (2017). The Sustainable Development Goals and the systems approach to sustainability. *Economics, 11*(1), 20170028.

Botai, J. O., Botai, C. M., Ncongwane, K. P., Mpandeli, S., Nhamo, L., Masinde, M., & Mabhaudhi, T. (2021). A review of the water–energy–food nexus research in Africa. *Sustainability, 13*(4), 1762.

Cerdà, V., Estela, J. M., Forteza, R., Cladera, A., Becerra, E., Altimira, P., & Sitjar, P. (1999). Flow techniques in water analysis. *Talanta, 50*(4), 695–705.

Chang, M., Thellufsen, J. Z., Zakeri, B., Pickering, B., Pfenninger, S., Lund, H., & Østergaard, P. A. (2021). Trends in tools and approaches for modelling the energy transition. *Applied Energy, 290*, Article 116731.

Chen, S., & Chen, B. (2014). Energy efficiency and sustainability of complex biogas systems: A 3-level emergetic evaluation. *Applied Energy, 115*, 151–163.

Chen, S., & Chen, B. (2015). Urban energy consumption: Different insights from energy flow analysis, input–output analysis and ecological network analysis. *Applied Energy, 138*, 99–107.

Chen, S., Wei, Z., Sun, G., Cheung, K. W., & Sun, Y. (2016). Multi-linear probabilistic energy flow analysis of integrated electrical and natural-gas systems. *IEEE Transactions on Power Systems, 32*(3), 1970–1979.

Correa-Posada, C. M., & Sanchez-Martin, P. (2014). Integrated power and natural gas model for energy adequacy in short-term operation. *IEEE Transactions on Power Systems, 30*(6), 3347–3355.

Costanza, R., Daly, L., Fioramonti, L., Giovannini, E., Kubiszewskia, I., Mortensen, L.F., Pickett, K.E., Ragnarsdottir, K., De Vogli, R., Wilkinson, R. (2016). Modelling and measuring sustainable wellbeing in connection with the UN Sustainable Development Goals. *Ecological Economics*, 130, 350–355.http://www.sciencedirect.com/science/article/pii/S09218009 15303359.

Dahl, A. L. (2012). Achievements and gaps in indicators for sustainability. *Ecological Indicators, 17*, 14–19.

Dai, J., Wu, S., Han, G., Weinberg, J., Xie, X., Wu, X., & Yang, Q. (2018). Water-energy nexus: A review of methods and tools for macro-assessment. *Applied Energy, 210*, 393–408.

Darko, D., Osaliya, R., Aziz, F., & Wekesa, D. (2023). Governing the nexus: Water-energy-food nexus governance strategies in Ghana and Uganda. *Environmental Development, 48*, Article 100933.

Davidsdottir, B., Ásgeirsson, E. I., Fazeli, R., Gunnarsdottir, I., Leaver, J., Shafiei, E., & Stefánsson, H. (2024). Integrated energy systems modelling with multi-criteria decision analysis and stakeholder engagement for identifying a sustainable energy transition. *Energies, 17*(17).

Davies, G., Prpich, G., Strachan, N., & Pollard, S. (2014). *Ukerc energy strategy under uncertainties: Identifying techniques for managing uncertainty in the energy sector.* UKERC Working Paper UKERC/WP/FG/2014/001.

De Loë, R. C., & Patterson, J. J. (2017). Rethinking water governance: Moving beyond water-centric perspectives in a connected and changing world. *Natural Resources Journal, 57*(1), 75–100.

De Strasser, L., Lipponen, A., Howells, M., Stec, S., & Bréthaut, C. (2016). A methodology to assess the water, energy food ecosystems nexus in transboundary river basins. *Water, 8*(2), 59.

Del Granado, P. C., Van Nieuwkoop, R. H., Kardakos, E. G., & Schaffner, C. (2018). Modelling the energy transition: A nexus of energy system and economic models. *Energy Strategy Reviews, 20*, 229–235.

Eberling, E., & Langkau, S. (2024). Utilising SDGs in sustainability assessments of innovations: Deriving methodological recommendations from existing approaches. *Journal of Cleaner Production, 437*, Article 140383.

Elmqvist, T., Cornell, S., Öhman, M. C., Daw, T., Moberg, F., Norström, A., & Ituarte-Lima, C. (2014). *Global Sustainability & Human Prosperity: Contribution to the post-2015 Agenda and the development of Sustainable Development Goals.* Nordic Council of Ministers.

Emina, K. A. (2021). Sustainable development and the future generations. *Social Sciences, Humanities and Education Journal (SHE Journal), 2*(1), 57–71.

Endo, A., Burnett, K., Orencio, P. M., Kumazawa, T., Wada, C. A., Ishii, A., & Taniguchi, M. (2015). Methods of the water-energy-food nexus. *Water, 7*(10), 5806–5830.

Endo, A., Tsurita, I., Burnett, K., & Orencio, P. M. (2017). A review of the current state of research on the water, energy, and food nexus. *Journal of Hydrology: Regional Studies, 11*, 20–30.

Endo, A., Yamada, M., Miyashita, Y., Sugimoto, R., Ishii, A., Nishijima, J., & Qi, J. (2020). Dynamics of water–energy–food nexus methodology, methods, and tools. *Current Opinion in Environmental Science & Health, 13*, 46–60.

Estoque, R. C. (2023). Complexity and diversity of nexuses: A review of the nexus approach in the sustainability context. *Science of the Total Environment, 854*, Article 158612.

Farmandeh, E., Choobchian, S., & Karami, S. (2024). Conducting water-energy-food nexus studies: What, why, and how. *Scientific Reports, 14*(1), 27310.

Fernandes Torres, C. J., Peixoto de Lima, C. H., Suzart de Almeida Goodwin, B., Rebello de Aguiar Junior, T., Sousa Fontes, A., Veras Ribeiro, D., & Dantas Pinto Medeiros, Y. (2019). A literature review to propose a systematic procedure to develop "nexus thinking" considering the water–energy–food nexus. *Sustainability, 11*(24), 7205.

Foran, T. (2015). Node and regime: Interdisciplinary analysis of water-energy-food nexus in the Mekong region. *Water Alternatives, 8*(1), 655–674.

Francis, K., Dongying, S., Dennis, A., Edmund, N. N. K., & Kumah, N. Y. G. (2022). Network governance and renewable energy transition in sub-Saharan Africa: Contextual evidence from Ghana. *Energy for Sustainable Development, 69*, 202–210.

Geels, F. W. (2002). Technological transitions as evolutionary reconfiguration processes: A multi-level perspective and a case study. *Research Policy, 31*(8–9), 1257–1274.

Geels, F. W. (2004). From sectoral systems of innovation to socio-technical systems: Insights about dynamics and change from sociology and institutional theory. *Research Policy, 33*(6–7), 897–920.

Geels, F. W. (2005). The dynamics of transitions in socio-technical systems: A multi-level analysis of the transition pathway from horse-drawn carriages to automobiles (1860–1930). *Technology Analysis & Strategic Management, 17*(4), 445–476.

Geels, F. W. (2011). The multi-level perspective on sustainability transitions: Responses to seven criticisms. *Environmental Innovation and Societal Transitions, 1*(1), 24–40.

Geels, F. W. (2018). Disruption and low-carbon system transformation: Progress and new challenges in socio-technical transitions research and the multi-level perspective. *Energy Research & Social Science, 37*, 224–231.

Geels, F. W. (2019). Socio-technical transitions to sustainability: A review of criticisms and elaborations of the multi-level perspective. *Current Opinion in Environmental Sustainability, 39*, 187–201.

Geels, F. W., & Schot, J. (2007). Typology of sociotechnical transition pathways. *Research Policy, 36*(3), 399–417.

Hamiche, A. M., Stambouli, A. B., & Flazi, S. (2016). A review of the water-energy nexus. *Renewable and Sustainable Energy Reviews, 65*, 319–331.

Harwood, S. A. (2018). In search of a (WEF) nexus approach. *Environmental Science & Policy, 83*, 79–85.

Hoff, H. (2011). *Understanding the nexus. Background paper for the Bonn 2011 conference: The water, energy and food security nexus.* Stockholm Environment Institute (SEI).

Jackson, M. C. (2024). *Critical systems thinking: A practitioner's guide.* Wiley.

Jain, P., & Jain, P. (2020). Are the Sustainable Development Goals really sustainable? A Policy Perspective. *Sustainable Development, 28*(6), 1642–1651.

Kullmann, F., Markewitz, P., Stolten, D., & Robinius, M. (2021). Combining the worlds of energy systems and material flow analysis: A review. *Energy, Sustainability and Society, 11*(1), 13.

Kumar, N., & Karmakar, S. (2024). Techno-economic optimisation of hydrogen generation through hybrid energy system: A step towards sustainable development. *International Journal of Hydrogen Energy, 55*, 400–413.

Kusakana, K. (2014). Techno-economic analysis of off-grid hydrokinetic-based hybrid energy systems for onshore/remote area in South Africa. *Energy, 68,* 947–957.

Leck, H., Conway, D., Bradshaw, M., & Rees, J. (2015). Tracing the water–energy–food nexus: Description, theory and practice. *Geography Compass, 9*(8), 445–460.

Li, H., Hou, K., Xu, X., Jia, H., Zhu, L., & Mu, Y. (2022). Probabilistic energy flow calculation for regional integrated energy system considering cross-system failures. *Applied Energy, 308,* Article 118326.

Littig, B., & Griessler, E. (2005). Social sustainability: A catchword between political pragmatism and social theory. *International Journal of Sustainable Development, 8*(1–2), 65–79.

Ma, W., Fang, S., Liu, G., & Zhou, R. (2017). Modelling of district load forecasting for distributed energy system. *Applied Energy, 204,* 181–205.

Mahlatsi, K. (2021). Achieving sustainable development goals (SDGs) in Africa: Challenges and prospects. *The Thinker, 87*(2), 61–69.

Markard, J., Raven, R., & Truffer, B. (2012). Sustainability transitions: An emerging field of research and its prospects. *Research Policy, 41*(6), 955–967.

Meadowcroft, J. (2007). Who is in charge here? Governance for sustainable development in a complex world. *Journal of Environmental Policy & Planning, 9*(3–4), 299–314, 314.

Mei, F., Zhang, J., Lu, J., Lu, J., Jiang, Y., Gu, J., & Gan, L. (2021). Stochastic optimal operation model for a distributed integrated energy system based on multiple-scenario simulations. *Energy, 219,* Article 119629.

Merton, R. K. (1968). *Social theory and social structure.* The Free Press.

Murphy, K. (2012). The social pillar of sustainable development: A literature review and framework for policy analysis sustainability. *Science, Practice and Policy, 8*(1), 15–29.

Oates, L. (2021). Sustainability transitions in the Global South: A multi-level perspective on urban service delivery. *Regional Studies, Regional Science, 8*(1), 426–433.

Pahl-Wostl, C. (2019). Governance of the water-energy-food security nexus: A multi-level coordination challenge. *Environmental Science & Policy, 92,* 356–367.

Paoli, A. D., & Addeo, F. (2019). Assessing SDGs: A methodology to measure sustainability. *Athens Journal of Social Sciences, 6*(3), 229–250.

Power, M., Newell, P., Baker, L., Bulkeley, H., Kirshner, J., & Smith, A. (2016). The political economy of energy transitions in Mozambique and South

Africa: The role of the rising powers. *Energy Research & Social Science, 17*, 10–19.

Pye, S., Sabio, N., & Strachan, N. (2015). An integrated systematic analysis of uncertainties in UK energy transition pathways. *Energy Policy, 87*, 673–684.

Rasul, G., & Sharma, B. (2016). The nexus approach to water–energy–food security: An option for adaptation to climate change. *Climate Policy, 16*(6), 682–702.

Rip, A., & Kemp, R. (1998). Technological change. In *Human choice and climate change: Vol. II, resources and technology* (pp. 327–399). Battelle Press.

Rogelj, J., Luderer, G., Pietzcker, R. C., Kriegler, E., Schaeffer, M., Krey, V., & Riahi, K. (2015). Energy system transformations for limiting end-of-century warming to below 1.5 °C. *Nature Climate Change, 5*(6), 519–527.

Sala, S., Ciuffo, B., & Nijkamp, P. (2015). A systemic framework for sustainability assessment. *Ecological Economics, 119*, 314–325.

Sampaio, H. C., Dias, R. A., & Balestieri, J. A. P. (2013). Sustainable urban energy planning: The case study of a tropical city. *Applied Energy, 104*, 924–935.

Schot, J., & Geels, F. W. (2013). Strategic niche management and sustainable innovation journeys: theory, findings, research agenda, and policy. *The Dynamics of Sustainable Innovation Journeys*, 17–34.

Schwanen, T. (2018). Thinking complex interconnections: Transition, nexus and Geography. *Transactions of the Institute of British Geographers, 43*(2), 262–283.

Shannak, S., Mabrey, D., & Vittorio, M. (2018). Moving from theory to practice in the water–energy–food nexus: An evaluation of existing models and frameworks. *Water-Energy Nexus, 1*(1), 17–25.

Siakwah, P., & Torto, O. (2022). Analysis of the complexities in the water-energy-food nexus: Ghana's Bui dam experience. *Frontiers in Sustainable Food Systems, 6*, Article 734675.

Simpson, G. B., & Jewitt, G. P. (2019). The development of the water-energy-food nexus as a framework for achieving resource security: A review. *Frontiers in Environmental Science, 7*, 8.

Smith, A., & Raven, R. (2012). What is protective space? Reconsidering niches in transitions to sustainability. *Research Policy, 41*(6), 1025–1036.

Smith, A., Stirling, A., & Berkhout, F. (2005). The governance of sustainable socio-technical transitions. *Research Policy, 34*(10), 1491–1510.

Sterl, S., Fadly, D., Liersch, S., Koch, H., & Thiery, W. (2021). Linking solar and wind power in eastern Africa with operation of the Grand Ethiopian Renaissance Dam. *Nature Energy, 6*(4), 407–418.

Stirling, A. (2015). *Developing 'Nexus Capabilities': Towards transdisciplinary methodologies*. University of Sussex. Retrieved from https://hdl.handle.net/10779/uos.23446508.v1.

Sungkawati, E. (2024). Opportunities and challenges: Adopting "blue-green economy" terms to achieve SDGs. *Revenue Journal: Management and Entrepreneurship, 2*(1), 01–13.

Swilling, M. (2006). Sustainability and infrastructure planning in South Africa: A Cape Town case study. *Environment and Urbanisation, 18*(1), 23–50.

Taşan-Kok, T., & Vranken, J. (2011). *Handbook for multilevel urban governance in Europe: Analysing participatory instruments for an integrated urban development*. European Urban Knowledge Network.

Unruh, G. C. (2000). Understanding carbon lock-in. *Energy Policy, 28*(12), 817–830.

Waughray, D. (Ed.). (2011). *Water security, the water-food-energy-climate nexus: The World Economic Forum Water Initiative*. Island Press.

WCED, S. W. S. (1987). World Commission on Environment and Development (WCED). *Our common future, 17*(1), 1–91.

Weitz, N., Nilsson, M., & Davis, M. (2014). A nexus approach to the post-2015 Agenda. *The SAIS Review of International Affairs, 34*(2), 37–50.

3

Demand Management Instruments for Water and Energy Security in the SDG Era

Robert C. Brears

Abstract This chapter examines demand management as a crucial strategy for attaining water and energy security in alignment with the Sustainable Development Goals (SDGs). It presents a comprehensive framework of economic, regulatory, informational, institutional, and technological instruments designed to reduce consumption, enhance efficiency, and shift usage patterns. Emphasizing cross-sectoral integration and digital innovation, the chapter highlights how demand-side approaches can optimize existing systems, improve resilience, and reduce environmental impacts. By aligning policy, planning, and behavioral tools, demand management offers a cost-effective pathway to more sustainable and equitable resource governance.

Keywords Demand management · Water-energy nexus · Resource efficiency · Sustainable Development Goals · Sustainable Infrastructure

© The Author(s), under exclusive license to Springer Nature Switzerland AG 2025
Y. Ermolaeva et al., *Rethinking Water and Energy for a Sustainable Future*, Palgrave Studies in Climate Resilient Societies, https://doi.org/10.1007/978-3-032-04485-3_3

Introduction

As global pressures on water and energy systems intensify due to population growth, urbanization, and climate change, managing demand has become a critical component of sustainable resource governance. Demand management focuses on reducing consumption, improving efficiency, and shifting usage patterns to alleviate pressure on supply systems and extend the life of infrastructure. Unlike supply-side strategies that emphasize resource expansion, demand-side approaches aim to optimize the use of existing resources, often resulting in lower environmental and financial costs.

In the context of the 2030 Agenda for Sustainable Development, demand management is directly linked to several Sustainable Development Goals (SDGs), notably SDG 6 (clean water and sanitation), SDG 7 (affordable and clean energy), SDG 12 (responsible consumption and production), and SDG 13 (climate action). These global goals highlight the importance of equitable access, efficiency, and resilience in managing essential services. Demand-side strategies play a crucial role in achieving these objectives by optimizing both the quantity and timing of resource use across sectors.

This chapter provides a structured overview of the main categories of demand management instruments, including economic, regulatory, informational, behavioral, institutional, and planning approaches, as well as digital and technological enablers. Each section outlines the function and rationale of the instruments, highlighting their contributions to reducing demand, enhancing system performance, and supporting the implementation of SDGs. Together, these instruments form a comprehensive toolkit for policymakers, utilities, and practitioners working to build more sustainable and efficient water and energy systems. Throughout the chapter, the relevance of each instrument is considered in terms of its contribution to achieving key SDGs.

Demand Management and the Water-Energy Nexus

Demand management plays a critical role in enhancing the sustainability and resilience of water and energy systems, particularly as global pressures on these resources intensify. The water-energy nexus underscores the interdependence between water and energy: water is essential for energy production and cooling processes, while energy is crucial for water treatment, distribution, and wastewater management. Demand-side strategies that reduce consumption or shift usage patterns can simultaneously ease pressures on both systems, offering co-benefits across sectors.

Traditional supply-side approaches have focused on expanding infrastructure and increasing resource extraction, but these often entail high financial and environmental costs. In contrast, demand management seeks to optimize existing systems by influencing consumer behavior and improving efficiency. In the context of the SDGs, demand management contributes to achieving SDG 6 (clean water and sanitation) and SDG 7 (affordable and clean energy), while also supporting SDG 12 (responsible consumption and production) and SDG 13 (climate action).

Integrated demand management requires coordinated planning and policy frameworks that address cross-sectoral linkages. For example, water-saving technologies in households can reduce the energy used for heating water, while energy-efficient irrigation practices in agriculture can reduce water withdrawals and energy inputs. Utility-led demand-side management (DSM) programs and national-level regulatory instruments can further embed efficiency into planning processes.

Digital technologies and data analytics enhance the effectiveness of demand management by enabling real-time monitoring, targeted interventions, and improved forecasting. Smart meters, consumption feedback tools, and automated controls offer consumers and providers greater visibility and control over resource use. These tools help identify inefficiencies and encourage behavioral changes that align with conservation goals. As climate variability increases, demand management becomes a vital tool for enhancing system flexibility and resilience, reducing reliance on supply augmentation, and supporting long-term resource sustainability.[1,2,3,4]

Economic Instruments

Economic instruments are key tools in demand management strategies, providing financial signals that influence consumer and institutional behavior toward more efficient use of water and energy. By assigning value to resources and externalities, these instruments help internalize environmental and social costs, encouraging users to conserve, shift consumption patterns, or invest in efficiency improvements. When well-designed, economic instruments can drive cost-effective demand reductions while maintaining service equity and supporting broader sustainability goals.

Tiered Pricing Structures

Tiered pricing structures are widely used as a demand management tool to encourage more efficient use of water and energy. These structures involve setting different unit prices for different levels of consumption, with higher usage levels charged at incrementally higher rates. The aim is to send a clear price signal that promotes conservation among high-volume users while maintaining affordability for basic needs. This approach aligns resource pricing with consumption behavior, helping to internalize the environmental and operational costs associated with excessive resource use. Tiered pricing supports SDG 6 (clean water and sanitation) and SDG 7 (affordable and clean energy) by promoting efficient use and ensuring access to essential services through lifeline rates.

In the water sector, increasing block tariffs typically apply lower rates to initial blocks of consumption to cover essential household needs, while applying higher rates to subsequent blocks associated with discretionary or excessive use. This pricing structure aims to strike a balance between equity and efficiency, protecting low-income households while promoting responsible consumption. In energy systems, tiered electricity pricing can serve a similar function, especially in residential sectors where peak demand drives infrastructure and generation costs. By linking higher prices to higher usage, utilities can influence demand patterns and reduce strain on supply systems during periods of high consumption. For tiered pricing to be effective, it must be supported by accurate metering, clear billing systems, and public understanding of how charges are applied. Poorly communicated or overly complex pricing tiers can lead to confusion or a perception of unfairness, thereby limiting their effectiveness. Additionally, tiered pricing must be carefully designed to avoid unintended consequences, such as penalizing larger households or encouraging the underuse of essential services. Complementary measures, such as lifeline rates for vulnerable populations or public education campaigns, can help ensure tiered pricing achieves its intended objectives while maintaining social acceptability. Properly implemented, tiered pricing can support the financial sustainability of utilities, reduce consumption levels, and contribute to more resilient water and energy systems.[5,6,7,8]

Performance-Based Rebates and Subsidies

Performance-based rebates and subsidies are economic instruments that incentivize users to adopt water- and energy-efficient technologies or practices. Unlike general subsidies, which are provided upfront or without verifying outcomes, these schemes tie financial support to demonstrated savings or efficiency improvements. By linking public or utility spending to proven conservation results, they enhance the cost-effectiveness of demand management efforts. These instruments support SDG 12 (responsible consumption and production) by promoting the uptake of resource-efficient technologies, while also contributing to SDG 6 and SDG 7 by reducing demand on water and energy systems through verified efficiency gains.

In the water sector, rebates may be offered to households or businesses for installing efficient appliances such as low-flow toilets, water-efficient irrigation systems, or greywater reuse technologies. The rebate amount depends on the volume of water saved, as verified through audits or consumption data. Similarly, in energy systems, subsidies for efficient lighting, heating, or cooling systems can be tied to actual reductions in energy use or peak demand. Performance-based models help ensure that incentives target actions with verifiable impacts, rather than simply subsidizing technology adoption regardless of performance.

To be effective, these schemes require robust monitoring and verification frameworks to assess performance. This includes baseline assessments, usage tracking, and post-installation evaluations. While such requirements can increase administrative complexity and costs, they improve the reliability and transparency of the programs. Explicit criteria for eligibility, standardized measurement protocols, and accessible application processes are essential for maximizing participation and achieving resource efficiency.

Equity considerations are also crucial in designing performance-based incentives. Without safeguards, these programs may disproportionately benefit higher-income households with the resources to make upfront investments. Solutions include providing enhanced incentives for low-income users or offering on-bill financing options that reduce financial barriers. When paired with effective communication and technical support, performance-based rebates and subsidies can encourage the adoption of efficiency measures, reduce long-term operating costs, and contribute to the broader objectives of sustainable water and energy management.[9,10]

Water and Energy Taxes

Water and energy taxes are fiscal instruments designed to influence consumption patterns by incorporating the environmental and resource costs of use into prices. By increasing the cost of water or energy use through taxation, these instruments create economic incentives for consumers and businesses to reduce demand, invest in efficiency, and shift toward more sustainable practices. Unlike subsidies or rebates, taxes serve as disincentives, aiming to internalize the negative externalities associated with the overuse, pollution, or depletion of resources. These instruments contribute to SDG 13 (climate action) by discouraging resource-intensive practices, and to SDG 6 and SDG 7 by promoting conservation and more sustainable water and energy use through pricing that reflects actual environmental costs.

In the water sector, abstraction taxes can be levied on the volume of water withdrawn from natural sources, encouraging conservation and more efficient allocation, particularly in areas experiencing scarcity. Wastewater discharge taxes may also be imposed to reduce pollution and fund the development of treatment infrastructure. In the energy sector, taxes can be applied to fossil fuels or electricity consumption to reflect environmental impacts, such as greenhouse gas emissions or air pollution. These taxes can vary by fuel type or usage level to reflect relative environmental burdens and policy priorities.

The effectiveness of water and energy taxes depends on several design factors, including the tax rate, structure, and scope of application. A well-calibrated tax must be high enough to influence behavior but not so high as to cause undue hardship, especially for lower-income users. Revenue generated can be recycled into programs that support vulnerable populations, fund infrastructure improvements, or invest in renewable energy and conservation initiatives. Transparency in tax use and clear communication about objectives can increase public acceptance and compliance.

Administrative capacity is essential for implementing and managing tax systems. This includes metering infrastructure, billing systems, and monitoring mechanisms to track consumption and ensure accountability. Taxes can be implemented at various governance levels, national, regional, or local, depending on policy objectives and institutional frameworks, offering flexibility in addressing diverse water and energy challenges.[11,12]

Regulatory Instruments

Regulatory instruments form a foundational component of demand management by setting mandatory standards and rules that govern water and energy use. These instruments establish minimum performance requirements, usage restrictions, or design specifications to ensure efficiency and reduce unnecessary consumption. Unlike economic or informational tools that rely on market signals or voluntary behavior, regulatory approaches mandate compliance, providing a direct and enforceable means of influencing consumption patterns across sectors and user groups.

Mandatory Efficiency Standards

Mandatory efficiency standards are regulatory tools that require products, systems, or practices to meet predefined levels of performance in terms of water or energy use. These standards are typically applied to appliances, equipment, buildings, and industrial processes to ensure a baseline level of efficiency across markets. By setting minimum performance requirements, governments can eliminate the least efficient options from the market and drive innovation toward more sustainable technologies. These standards directly support SDG 7 (affordable and clean energy) and SDG 6 (clean water and sanitation) by raising minimum efficiency levels across sectors, while contributing to SDG 13 (climate action) through reduced emissions and lower resource demand.

In the water sector, efficiency standards often apply to fixtures such as toilets, taps, showerheads, and irrigation equipment. These standards specify maximum flow rates or consumption per use, thereby reducing overall water demand without relying on changes in user behavior. In the energy sector, mandatory standards cover a wide range of appliances, including lighting, refrigerators, air conditioners, and industrial motors. These standards are periodically updated to reflect technological advances and changing policy targets, contributing to incremental improvements in efficiency over time.

The effectiveness of mandatory standards depends on the development of appropriate testing procedures, certification processes, and enforcement mechanisms. Manufacturers must be able to demonstrate compliance through standardized performance testing, while governments need to have systems in place to monitor products on the market and penalize non-compliance. Clear labeling schemes can complement standards by helping consumers identify compliant products.

Mandatory standards can be designed to apply at different levels, product, building, or system, depending on the policy objective. For example, building codes can incorporate mandatory water and energy efficiency criteria for new construction or significant renovations. In industrial settings, performance benchmarks may apply to entire processes or facilities. While initial compliance may involve higher upfront costs for manufacturers or developers, these are often offset by long-term savings for consumers and reduced demand on infrastructure. Mandatory efficiency standards serve as a baseline upon which other demand management strategies can be built, ensuring minimum levels of sustainability across the system.[13,14,15,16]

Building and Plumbing Codes

Building and plumbing codes are regulatory instruments that set design and construction requirements to promote water and energy efficiency in the built environment. These codes establish technical standards for systems such as lighting, insulation, heating, cooling, ventilation, and water delivery and drainage. By embedding efficiency measures into building design and construction practices, codes help reduce long-term resource consumption and operational costs across residential, commercial, and industrial sectors. These codes contribute to SDG 11 (sustainable cities and communities), SDG 6 (clean water and sanitation), and SDG 7 (affordable and clean energy) by ensuring that new and existing buildings operate efficiently, reduce resource demand, and support more sustainable urban development.

In the water sector, plumbing codes regulate the design and installation of water supply and sanitation systems. They often specify maximum allowable flow rates, pipe sizing, and fixture efficiency, such as limiting the water used per flush in toilets or per minute in showerheads. These requirements ensure that new or renovated buildings incorporate water-saving technologies as a default, reducing the reliance on individual consumer behavior to drive efficiency.

For energy systems, building codes may address insulation levels, glazing standards, air leakage limits, and the integration of efficient heating and cooling systems. Advanced codes can also mandate the use of passive design principles, solar-ready infrastructure, or minimum renewable energy contributions. As energy performance in buildings directly affects peak demand and emissions, codes are a key mechanism for improving the sustainability of urban development.

The enforcement of building and plumbing codes typically falls under the jurisdiction of local or regional authorities. Compliance is verified through design approvals, inspections, and certification processes during construction and occupancy phases. Regular code updates are necessary to reflect technological advancements and evolving policy targets, requiring collaboration between regulatory bodies, industry stakeholders, and technical experts. Building and plumbing codes complement other demand management tools by ensuring that new developments start from a baseline of efficiency, embedding conservation into the structure and function of the built environment over its lifecycle.[17,18,19]

Time-of-Use Restrictions or Caps

Time-of-use restrictions and caps are regulatory measures designed to influence when and how much water or energy is consumed, particularly during periods of high demand or limited supply. These instruments aim to shift consumption away from peak periods, thereby reducing stress on infrastructure, improving system reliability, and optimizing resource use. By imposing limits or restrictions during specific times of day or under particular conditions, they help manage demand in real-time and support more balanced load profiles. These measures support SDG 7 (affordable and clean energy) and SDG 6 (clean water and sanitation) by promoting efficient, demand-aligned consumption and contribute to SDG 13 (climate action) through reduced peak loads and associated emissions.

In the energy sector, time-of-use regulations may restrict the use of certain appliances during peak hours or impose caps on total consumption within a defined period. These restrictions are often paired with variable pricing or smart technologies to reinforce behavioral adjustments. For example, air conditioning or electric vehicle charging may be limited or scheduled outside of high-demand windows to prevent grid congestion or reduce the need for additional generation capacity.

In the water sector, restrictions can include limits on outdoor water use, such as lawn irrigation or car washing, during times of drought or peak demand. Some jurisdictions implement seasonal or time-of-day caps to align water use with supply availability, particularly in regions that rely on surface water or experience variable rainfall. These measures are usually communicated through public notices and enforced through monitoring and penalties for non-compliance.

The success o time-of-use restrictions and caps depends on public awareness, clear communication, and enforceability. Complementary tools such as automated timers, smart meters, and usage alerts can facilitate compliance and reduce inconvenience for users. Regulatory authorities must balance the need to manage demand with maintaining fairness and accessibility, particularly for users with limited flexibility in when they consume resources. When implemented effectively, time-based controls offer a flexible and responsive approach to demand management that addresses both supply constraints and system efficiency.[20,21,22]

Informational and Behavioral Instruments

Informational and behavioral instruments are designed to influence water and energy use by targeting user awareness, perceptions, and decision-making processes. Rather than relying on financial incentives or regulatory mandates, these tools aim to shift behavior through the provision of information, feedback, and subtle cues that encourage more efficient consumption practices. Rooted in behavioral science and communication strategies, these instruments play a complementary role in demand management by addressing the psychological and social drivers of consumer behavior.

Smart Metering and Feedback Systems

Smart metering and feedback systems are informational tools that support real-time or near real-time monitoring of water and energy consumption. They provide users with detailed data on their consumption habits, enabling them to make more informed decisions and make timely adjustments to their behavior. For utilities, these systems enhance data accuracy, improve operational efficiency, and help identify issues such as leaks, tampering, or abnormal demand. By promoting transparency and enabling responsive management, they contribute to SDG 12 (responsible consumption and production), while also advancing SDG 6 and SDG 7 through improved efficiency and service reliability in water and energy systems.

Smart meters record consumption at frequent intervals and transmit data to utilities and consumers via digital communication networks. This contrasts with traditional meters, which require manual readings and provide limited data. By offering granular usage information, such as hourly or daily consumption, smart meters empower users to understand how and when they use resources and to identify opportunities for efficiency improvements. In many cases, users receive feedback through in-home displays, web portals, or mobile applications that visualize usage trends and suggest actions to reduce demand.

Feedback systems can be designed to include comparative metrics, such as benchmarking against past performance or similar households. These comparisons can encourage conservation by highlighting excessive usage or reinforcing efficient behavior. Some systems also offer alerts for threshold breaches, budget tracking, or goal-setting features that enhance user engagement.

For utilities, smart metering supports demand forecasting, load management, and customer segmentation. The data generated enables targeted communication and the design of personalized efficiency programs. In water systems, early leak detection can prevent losses and reduce strain on infrastructure. In energy systems, consumption data can inform demand response programs that adjust usage based on grid conditions.

The successful implementation of smart metering and feedback systems relies on robust data privacy protections, user-friendly interfaces, and public trust in technology. When effectively deployed, these systems improve transparency, enhance user participation in demand management, and contribute to more resilient and efficient utility operations.[23,24,25,26]

Water and Energy Labeling and Certification Schemes

Water and energy labeling and certification schemes are informational tools that guide consumer choices by providing clear, standardized indicators of a product's resource efficiency. These schemes aim to make the efficiency performance of appliances, equipment, and buildings transparent at the point of purchase or use, enabling users to consider long-term operating costs and environmental impacts alongside the upfront price. By encouraging demand for efficient products, labeling and certification contribute to market transformation and reinforce other demand management measures. These schemes support SDG 12 (responsible consumption and production) by facilitating informed purchasing decisions and contribute to SDG 6 and SDG 7 by promoting widespread adoption of water- and energy-efficient technologies.

Labels typically display information such as efficiency ratings, estimated annual resource use, or performance classifications on a simple visual scale. Standard formats include color-coded bars or star systems, which facilitate easy comparison between models. Certification schemes often go further, providing formal recognition that a product or building meets specific efficiency criteria or sustainability standards. These may involve third-party verification and are frequently linked to broader environmental rating systems.

In the water sector, labeling schemes apply to fixtures such as toilets, showerheads, taps, and washing machines, indicating their flow rates or water consumption per cycle. In the energy sector, labels are used for appliances, lighting, heating and cooling systems, and are increasingly applied to entire buildings. Labels can drive innovation by encouraging manufacturers to enhance product performance, meeting or exceeding efficiency benchmarks.

The credibility and effectiveness of labeling and certification schemes depend on standardization, clarity, and enforcement. Programs must be based on consistent testing protocols and regularly updated to reflect technological advancements. Consumer education is also essential to ensure that labels are understood and used in purchasing decisions. When integrated into procurement policies, incentives, or building codes, labeling and certification schemes can support broader policy objectives and accelerate the adoption of resource-efficient technologies across residential, commercial, and industrial sectors.[27,28,29,30]

Public Awareness Campaigns and Behavioral Nudges

Public awareness campaigns and behavioral nudges are tools used to influence water and energy use by promoting changes in individual and collective behavior. Their goal is to increase understanding, reshape attitudes, and prompt specific actions by raising awareness of consumption habits and the consequences of inefficiency. While campaigns focus on broad communication and education, nudges aim to subtly influence decision-making environments, promoting more resource-conscious choices without limiting freedom of choice. These instruments support SDG 12 (responsible consumption and production) by encouraging sustainable behavior and contribute to SDG 6 and SDG 7 by highlighting the value and efficient use of water and energy.

Awareness campaigns typically rely on mass media, community outreach, and educational initiatives to share information on water and energy challenges. They may communicate the environmental and financial costs of overuse, offer conservation tips, or promote incentives and technologies. Effective messaging is tailored to target audiences, uses relatable language, and often combines factual content with emotional or visual appeal to engage the public.

Behavioral nudges draw on the principles of behavioral economics and psychology to encourage efficient resource use by modifying the context in which decisions are made. Examples include placing visual reminders near taps or switches, using default settings that favor efficiency, or comparing individual consumption to that of peers on utility bills. These low-cost interventions are designed to make sustainable choices more accessible, intuitive, or socially reinforced.

The success of these approaches depends on context specificity, consistency over time, and integration with other instruments such as pricing signals or regulations. They are especially effective in reaching segments of the population that are less responsive to economic or legal incentives. Measuring their impact involves tracking changes in usage and evaluating how particular messages or interventions affect behavior. When thoughtfully designed and deployed, public awareness campaigns and behavioral nudges can reinforce broader demand management strategies and help drive lasting behavioral change.[31,32,33,34]

Institutional and Planning Instruments

Institutional and planning instruments provide the structural and procedural foundation for implementing demand management in water and energy systems. These tools operate at the organizational and policy levels, enabling the integration of efficiency goals into long-term planning, regulatory mandates, and cross-sectoral coordination efforts. Unlike instruments focused on individual consumers or specific technologies, institutional and planning measures shape how decisions are made, resources are allocated, and responsibilities are shared across agencies and sectors.

Integrated Resource Planning

Integrated Resource Planning (IRP) is a strategic framework that assesses both supply-side and demand-side options to meet future water and energy needs in a cost-effective, reliable, and environmentally sustainable way. By combining long-term demand forecasts with system constraints and policy goals, IRP enables the development of balanced, integrated solutions. Unlike traditional planning approaches that priorities supply expansion, the IRP assigns equal importance to demand management, recognizing its potential to deliver more sustainable and cost-effective outcomes. This approach directly supports SDG 6 (clean water and sanitation), SDG 7 (affordable and clean energy), and SDG 13 (climate action) by advancing forward-looking strategies that enhance efficiency, system resilience, and environmental integrity.

In the energy sector, integrated planning involves evaluating options for fossil fuel and renewable generation, storage technologies, and measures that reduce consumption, including energy efficiency upgrades, load shifting, and behavioral interventions. In water systems, it involves assessing supply options, such as desalination or groundwater extraction, alongside conservation, leakage reduction, and water reuse. By placing all options on an equal footing, the approach helps identify least-cost pathways that balance reliability with environmental performance.

The planning process typically features scenario modeling, stakeholder engagement, and sensitivity analysis to manage uncertainties in areas such as demand growth, climate impacts, and regulatory shifts. Evaluation criteria often go beyond financial cost to include social equity, emissions reduction, and ecological outcomes, enabling more comprehensive and resilient decision-making.

Institutional capacity plays a crucial role in the successful implementation of initiatives. Utilities and planning agencies must be equipped to collect and analyze data, model future system behavior, and incorporate diverse stakeholder perspectives. The resulting strategies are reflected in investment plans, tariff structures, and policy directions, ensuring that long-term priorities are embedded in operational practices. Periodic updates are essential to account for evolving technologies and shifting conditions, ensuring that demand-side measures are embedded in future planning cycles.[35,36,37,38]

Utility Demand Side Management Programs

Utility demand-side management (DSM) programs are structured initiatives implemented by water and energy utilities to influence customer consumption through efficiency improvements, behavioral change, and load management. These programs aim to reduce overall demand, shift usage to off-peak periods, or minimize system losses, thereby enhancing the operational efficiency and sustainability of utility services. By targeting consumption at the point of use, DSM complements supply-side investments. These efforts support SDG 6 (clean water and sanitation) and SDG 7 (affordable and clean energy) through more efficient resource utilization, while also contributing to SDG 13 (climate action) by reducing emissions and alleviating infrastructure strain.

In the energy sector, DSM measures often include incentives for efficient appliances, support for building retrofits, and demand response initiatives that reduce or shift load during peak times. For water utilities, strategies may involve rebates for efficient fixtures, leak detection and repair programs, and public education campaigns to promote conservation. Programs are typically designed for both residential and commercial users, with tailored interventions based on consumption patterns and system priorities.

Successful implementation requires access to detailed usage data, often made possible by smart meters, and the ability to segment customers based on their behavior and demand patterns. Program design usually involves market analysis, goal setting, cost–benefit evaluations, and robust monitoring. Financial tools such as rebates or time-of-use pricing are commonly employed to boost participation, complemented by outreach and technical support.

Regulatory frameworks and cost recovery mechanisms play a vital role in enabling utility investment in DSM. In many regions, utilities are either permitted or mandated to fund these programs as part of their service obligations, with costs recovered through tariffs or public funding. Outcomes are typically evaluated using metrics such as peak demand reduction, water or energy savings, and engagement levels. Well-designed DSM programs can postpone the need for new infrastructure, lower environmental impacts, and better align utility operations with sustainability and climate policy goals.[39,40,41,42]

Cross-Sector Coordination and Governance

Cross-sector coordination and governance are essential components of effective demand management, particularly given the interdependencies between water, energy, agriculture, and urban systems. These instruments aim to align policy objectives, regulatory frameworks, and operational practices across various sectors and levels of government, ensuring coherent and mutually reinforcing strategies. Without coordination, demand management efforts risk fragmentation, duplication, or unintended consequences that undermine overall resource efficiency. This coordination advances SDG 6 (clean water and sanitation), SDG 7 (affordable and clean energy), SDG 11 (sustainable cities and communities), and SDG 13 (climate action) by promoting integrated planning and governance that reflect the interconnected nature of resource systems.

Effective governance frameworks establish clear roles and responsibilities among institutions involved in water and energy planning and service delivery. This may include national ministries, regional authorities, municipal utilities, and regulatory agencies. Coordination mechanisms can take the form of inter-agency working groups, joint planning processes, shared data platforms, or integrated policy instruments. These arrangements facilitate the exchange of information, harmonies standards, and promote joint decision-making on the use of resources and infrastructure development.

At the operational level, cross-sector coordination supports the identification of synergies and trade-offs. For example, energy efficiency improvements in water pumping or treatment can reduce electricity demand, while water-saving technologies in agriculture can lower both water and energy inputs. Aligning goals across sectors enables more effective prioritization of interventions and prevents shifting burdens from one system to another.

Governance structures must also facilitate stakeholder participation, including that of the private sector, civil society, and end-users. Inclusive processes enhance transparency, foster trust, and enhance the acceptability and effectiveness of demand management policies. Institutional capacity, legal frameworks, and political commitment all influence the quality of coordination. In contexts where responsibilities are fragmented or overlapping, reforms may be necessary to establish more explicit mandates and improve accountability.

Overall, cross-sector coordination and governance mechanisms help integrate demand-side considerations into broader resource management strategies, ensuring that policies and actions reflect the complex, interconnected nature of water and energy systems.[43,44,45,46]

Digital and Technological Enablers

Digital and technological tools are playing a pivotal role in advancing demand management strategies across the water and energy sectors. These innovations enhance data collection, analysis, and decision-making, enabling utilities, policymakers, and consumers to monitor consumption in real time, detect inefficiencies, and adapt to changing demand conditions. By improving the precision and responsiveness of demand-side interventions, digital technologies contribute to more resilient and efficient resource systems. These enablers directly support SDG 6 (clean water and sanitation), SDG 7 (affordable and clean energy), and SDG 13 (climate action) by fostering data-driven, adaptive approaches that reduce resource losses, improve system performance, and enhance climate resilience.

Smart meters are among the most widely adopted digital solutions. They record water or energy usage at frequent intervals, generating high-resolution data that can be analyzed to detect peak loads, irregular consumption patterns, or system inefficiencies. This supports infrastructure planning, load forecasting, and targeted efficiency programs for utilities. For consumers, these devices provide visibility and control over resource use, particularly when paired with user-friendly interfaces, such as mobile apps or in-home displays.

Advanced data analytics and machine learning models further amplify the benefits of digital metering by facilitating leak detection, predictive infrastructure maintenance, and customer segmentation for tailored interventions. These capabilities help optimize operations, reduce non-revenue losses, and enhance the targeting of conservation programs. In the energy sector, digital platforms underpin demand response schemes by automating load adjustments based on grid conditions. In the water sector, automated pressure regulation and flow control systems help maintain equilibrium between supply and demand.

Digital twins, virtual replicas of physical infrastructure, are emerging as powerful tools for planning and management. By integrating sensor inputs, simulation models, and operational data, utilities can test scenarios, assess vulnerabilities, and refine strategies before implementation. Likewise, geographic information systems (GIS) are widely used to visualize resource distribution, evaluate service coverage, and identify areas that require intervention.

The Internet of Things (IoT) strengthens infrastructure connectivity by linking meters, valves, pumps, and other assets into intelligent networks. These networks support remote monitoring and system control, reduce manual workload, and facilitate rapid responses to operational changes. In off-grid or rural areas, IoT-enabled systems expand the reach of demand management by remotely monitoring usage and system performance.

Cloud computing underpins the storage and analysis of the large volumes of data generated by digital infrastructure, enabling utilities and regulators to scale analytical tools as needed. Open data platforms and visual dashboards enhance transparency, enable public participation, and support collaborative solutions. At the same time, blockchain is being explored for its potential to secure and decentralize energy and water trading systems, offering tamper-proof transaction records between producers and consumers.

To fully leverage these technologies, attention must be paid to data privacy, system interoperability, and capacity development. Ensuring that platforms can integrate across systems and that users can interpret and act on data insights is critical. Regulatory frameworks may also need to evolve to accommodate real-time data in areas such as planning, billing, and service delivery. As digital and technological enablers continue to mature, they are set to play an increasingly central role in advancing flexible, adaptive, and efficient demand management systems.[47,48,49,50]

Conclusion

Demand management is central to achieving the SDGs, shifting the focus from expanding supply to ensuring the efficient, equitable use of existing water and energy systems. This chapter outlines a diverse set of instruments that, both individually and collectively, facilitate the transition to more sustainable consumption patterns. Together, these tools directly advance SDG 6 (clean water and sanitation), SDG 7 (affordable and clean energy), SDG 12 (responsible consumption and production), and SDG 13 (climate action).

Economic instruments, such as tiered pricing, performance-based subsidies, and resource taxes, create incentives for users to reduce demand, adopt efficient technologies, and make more sustainable consumption choices. By reducing pressure on water and energy systems while ensuring access to essential services, these tools support the equity and sustainability objectives embedded in SDGs 6 and 7, while promoting responsible resource use in line with SDG 12.

Regulatory mechanisms, including efficiency standards, building codes, and time-of-use controls, embed conservation requirements into infrastructure design and daily operations. These instruments enable durable reductions in resource use and emissions, supporting more climate-resilient cities and systems aligned with the goals of SDGs 7 and 13.

Informational and behavioral approaches empower consumers to manage their demand actively. Real-time feedback, labeling schemes, public campaigns, and behavioral nudges all contribute to increased awareness and encourage sustainable habits. These tools are essential to driving individual and community-level shifts that underpin progress toward SDG 12.

Planning and institutional frameworks ensure that demand-side strategies are systematically integrated into governance and infrastructure development. Instruments such as integrated resource planning and utility-led demand-side management embed efficiency into long-term decision-making, while coordination across sectors enhances coherence and effectiveness. This supports the implementation of holistic strategies consistent with the SDG framework.

Digital and technological innovations enhance the effectiveness of all demand management efforts. Smart meters, IoT networks, data analytics, and digital platforms enable real-time monitoring, adaptive responses, and targeted interventions. These capabilities are essential to building responsive systems that can withstand growing environmental pressures, supporting SDG 13's emphasis on climate resilience.

Together, these instruments offer a comprehensive toolkit for advancing sustainable water and energy systems. Institutionalizing demand management in national policies, utility operations, and investment strategies can accelerate progress across multiple SDGs, especially those at the intersection of environmental stewardship, infrastructure efficiency, and equitable service delivery.

Notes

1. R.C. Brears, *The Green Economy and the Water-Energy-Food Nexus*, Second ed. (Cham, Switzerland: Springer International Publishing, 2023).
2. *Urban Water Security* (Chichester, UK; Hoboken, NJ: John Wiley & Sons, 2016).
3. UN-Water, "Water, Food and Energy," https://www.unwater.org/water-facts/water-food-and-energy.
4. Redha Agadi et al., "Integration of Renewable Energy Resources into the Water-Energy-Food Nexus–Modeling a Demand Side Management Approach and Application to a Microgrid Farm in Morocco," *Frontiers in Environmental Economics* Volume 2—2023 (2023).
5. World Bank, "Water: A Scorecard for India (Tariffs & Subsidies in South Asia," (2022), https://documents1.worldbank.org/curated/en/270761468781549442/pdf/265390PAPER0WSP0Water0tariffs0no-02.pdf.
6. IEA, "Energy Prices," (2024), https://iea.blob.core.windows.net/assets/2c6e19c9-b62c-4116-83a7-283a25576f50/EnergyPrices Documentation.pdf.

7. Steven M. Smith, "The Effects of Individualized Water Rates on Use and Equity," *Journal of Environmental Economics and Management* 114 (2022).
8. Mohammad Ansarin et al., "A Review of Equity in Electricity Tariffs in the Renewable Energy Era," *Renewable and Sustainable Energy Reviews* 161 (2022).
9. Pennsylvania State University, "Types of Subsidies and Incentive Programs," https://www.e-education.psu.edu/eme801/node/504.
10. Brears, *The Green Economy and the Water-Energy-Food Nexus*.
11. Ibid.
12. Adam Tipper and Jane Harkness, "Environmental Taxation and Expenditure in New Zealand," (2018), https://www.wgtn.ac.nz/_ _data/assets/pdf_file/0004/1863211/WP10-Environmental-Tax ation-and-Expenditure-in-NZ.pdf.
13. UNFCCC, "Mandatory Performance Standards(E.G. Energy Performance Standard, Fuel Economy Standard, Renewable Portfolio Standard)," https://unfccc.int/policy/mandatory-perfor mance-standardseg-energy-performance-standard-fuel-economy-standard-renewable.
14. Elizabeth A. Kirk and Laurel Besco, "Improving Energy Efficiency: The Significance of Normativity," *Journal of Environmental Law* 33, no. 3 (2021).
15. Stephen Berry, Trivess Moore, and Michael Ambrose, "Australia's Experience of Combining Building Energy Standards and Disclosure Regulation," *Frontiers in Sustainable Cities* Volume 4—2022 (2022).
16. Brears, *The Green Economy and the Water-Energy-Food Nexus*.
17. J. Srinivasan, Orellana, E., Chiseilov, V., Guittonneau, C., & Doan, T., (2024), https://blogs.worldbank.org/en/developmentt alk/how-do-building-energy-codes-and-standards-measure-up--unveiling.
18. Troy William Heffernan et al., "The Carrot and the Stick: Policy Pathways to an Environmentally Sustainable Rental Housing Sector," *Energy Policy* 148 (2021).
19. Thomas E. Pape, "Plumbing Codes and Water Efficiency: What's a Water Utility to Do?," *Journal AWWA* 100, no. 5 (2008).

20. Presley K. Wesseh and Boqiang Lin, "A Time-of-Use Pricing Model of the Electricity Market Considering System Flexibility," *Energy Reports* 8 (2022).
21. Robert C. Brears, *Water Resources Management: Innovative and Green Solutions*, 2 ed. (Berlin, Boston: De Gruyter, 2024).
22. Brears, *Urban Water Security*.
23. *The Green Economy and the Water-Energy-Food Nexus.*
24. *Urban Water Security.*
25. *Regional Water Security* (Wiley, 2021).
26. Brears, *Water Resources Management: Innovative and Green Solutions.*
27. Brears, *The Green Economy and the Water-Energy-Food Nexus.*
28. *Urban Water Security.*
29. *Regional Water Security.*
30. Brears, *Water Resources Management: Innovative and Green Solutions.*
31. Brears, *The Green Economy and the Water-Energy-Food Nexus.*
32. *Urban Water Security.*
33. *Regional Water Security.*
34. Brears, *Water Resources Management: Innovative and Green Solutions.*
35. Brears, *The Green Economy and the Water-Energy-Food Nexus.*
36. *Urban Water Security.*
37. *Regional Water Security.*
38. Brears, *Water Resources Management: Innovative and Green Solutions.*
39. Brears, *The Green Economy and the Water-Energy-Food Nexus.*
40. *Urban Water Security.*
41. *Regional Water Security.*
42. Brears, *Water Resources Management: Innovative and Green Solutions.*
43. Brears, *The Green Economy and the Water-Energy-Food Nexus.*
44. *Urban Water Security.*
45. *Regional Water Security.*
46. Brears, *Water Resources Management: Innovative and Green Solutions.*

47. Brears, *The Green Economy and the Water-Energy-Food Nexus.*
48. *Urban Water Security.*
49. *Regional Water Security.*
50. Brears, *Water Resources Management: Innovative and Green Solutions.*

References

Agadi, R., Sakhraoui, K., Dupke, R. K. M., Wiebrow, E., & von Hirschhausen, C. (2023). Integration of renewable energy resources into the water-energy-food nexus–modeling a demand side management approach and application to a microgrid farm in Morocco. [In English]. *Frontiers in Environmental Economics, 2* (2023-August-24 2023).

Ansarin, M., Ghiassi-Farrokhfal, Y., Ketter, W., & Collins, J. (2022). A review of equity in electricity tariffs in the renewable energy era. *Renewable and Sustainable Energy Reviews, 161*, Article 112333.

Berry, S., Moore, T., & Ambrose, M. (2022). Australia's experience of combining building energy standards and disclosure regulation. [In English]. *Frontiers in Sustainable Cities, 4* (2022-January-27 2022).

Brears, R. C. (2016). *Urban water security.* Wiley.

Brears, R. C. (2021). *Regional water security.* Wiley.

Brears, R. C. (2023). *The green economy and the water-energy-food nexus* (2nd ed.). Springer.

Brears, R. C. (2024). *Water resources management: Innovative and green solutions* (2nd ed.). De Gruyter. https://doi.org/10.1515/9783111028101

Heffernan, T. W., Daly, M., Heffernan, E. E., & Reynolds, N. (2021). The Carrot and the stick: Policy pathways to an environmentally sustainable rental housing sector. *Energy Policy, 148*, Article 111939.

IEA. (2024) *Energy Prices.* (2024). Retrieved from https://iea.blob.core.windows.net/assets/2c6e19c9-b62c-4116-83a7-283a25576f50/EnergyPrices Documentation.pdf.

Kirk, E. A., & Besco, L. (2021). Improving energy efficiency: The significance of normativity. *Journal of Environmental Law, 33*(3), 669–695.

Pape, T. E. (2008). Plumbing codes and water efficiency: What's a water utility to do? *Journal AWWA, 100*(5), 101–103.

Pennsylvania State University. *Types of subsidies and incentive programs*. Retrieved from https://www.e-education.psu.edu/eme801/node/504.

Smith, S. M. (2022). The effects of individualized water rates on use and equity. *Journal of Environmental Economics and Management, 114*, Article 102673.

Srinivasan, J., Orellana, E., Chiseilov, V., Guittonneau, C., & Doan, T. (2024). *How do building energy codes and standards measure up? Unveiling a new global dataset*. World Bank.

Tipper, A., & Harkness, J. (2018). *Environmental taxation and expenditure in New Zealand*. Retrieved from https://www.wgtn.ac.nz/__data/assets/pdf_file/0004/1863211/WP10-Environmental-Taxation-and-Expenditure-in-NZ.pdf.

UN-Water. *Water, food and energy*. Retrieved from https://www.unwater.org/water-facts/water-food-and-energy.

UNFCCC. *Mandatory performance standards (E.G. energy performance standard, fuel economy standard, renewable portfolio standard)*. Retrieved from https://unfccc.int/policy/mandatory-performance-standardseg-energy-performance-standard-fuel-economy-standard-renewable.

Wesseh, P. K., & Lin, B. (2022). A time-of-use pricing model of the electricity market considering system flexibility. *Energy Reports, 8*, 1457–1470.

World Bank. (2022). *Water: A scorecard for India (Tariffs & Subsidies in South Asia)*. Retrieved from https://documents1.worldbank.org/curated/en/270761468781549442/pdf/265390PAPER0WSP0Water0tariffs0no-02.pdf.

4

Sustainable Marine and Offshore Energy Management

Yulia Ermolaeva

Abstract This chapter examines the role of water as a renewable energy resource, focusing on the technological, environmental, and socio-economic aspects and evaluation of the potential of the hydro and marine energy. It explores how sustainable practices in marine and offshore energy can address environmental challenges, benefit economies, and foster social well-being across different cultures. Firstly will be discussed global potential of hydro and marine energy, vulnerabilities barriers and principles of sustainable energy management with technological and socio-economic challenges on the path of implementation of the innovative technologies and practices from a cross-cultural viewpoint.

Keywords Marine energy · Hydro energy · Technological potential · Subject–object relationships · Energy transition · Energy management · Cross-cultural approach

© The Author(s), under exclusive license to Springer Nature Switzerland AG 2025
Y. Ermolaeva et al., *Rethinking Water and Energy for a Sustainable Future*, Palgrave Studies in Climate Resilient Societies, https://doi.org/10.1007/978-3-032-04485-3_4

Global Potential of Marine and Hydro-energy

The intersection of hydropower and marine energy systems with sustainable development encapsulates a multifaceted social dilemma: how to reconcile technological imperatives with equitable socio-ecological transitions. While these renewable energy sectors promise decarbonization and energy security, their deployment is deeply entangled with structural inequalities, governance paradigms, and contested notions of progress. Socio-technical approaches to this domain must interrogate the historical institutionalism shaping policy frameworks and energy management, the power asymmetries embedded in transnational energy projects, and the cultural-political dynamics of local communities affected by infrastructure. Conceptualizing hydropower and marine energy as sociotechnical systems reveals how technical innovations are inseparable from the social imaginaries that frame them, whether as tools of climate resilience or instruments of resource extraction. This analysis demands mixed-method frameworks: measuring structural impacts through lifecycle assessments, mapping subjectivities via stakeholder narratives, and historicizing institutional legacies that perpetuate or disrupt energy inequities.

Hydropower, as a cornerstone of renewable energy, exhibits profound regional disparities in utilization shaped by geographical, economic, and infrastructural factors. The global hydropower sector is a tapestry of contrasts marked by high utilization in geologically advantaged nations and untapped potential in resource-rich but governance-poor regions. Addressing these disparities demands tailored strategies: leveraging small-scale, ecological designs in sensitive areas, fostering international collaboration for mega-projects, and integrating climate resilience into planning.

In nations such as Norway, Switzerland, and France, over 90% of the economic hydropower potential is harnessed, a feat attributed to advanced infrastructure, mountainous topography, and cohesive policy frameworks (*Europe Hydropower Regional profile Hydropower in Europe*, 2025). Norway's fjord-rich landscape and Switzerland's Alpine rivers provide ideal conditions for high-efficiency plants, while France's centralized energy planning has prioritized hydropower as a stable renewable source. Conversely, countries like Russia, some regions in South Asia and South America lag significantly, utilizing only around 2–21% their potential. Russia's vast Siberian and Far Eastern river systems remain underdeveloped due to logistical complexities, funding shortages, and environmental opposition to large-scale projects (Mikhailov et al., 2021). Similarly, in some of developing countries, with a great theoretical capacity, is stymied by social-political instability, inadequate transmission infrastructure, and ecological concerns over biodiversity loss (Łącka, 2023).

Technological and environmental trade-offs further complicate global hydropower dynamics. The evaluation of technical potential for hydro and marine energy relies on hierarchical resource tiers and technology-specific indices (Satymov et al., 2024). The three-tier framework—distinguishing theoretical, technical, and practical potentials -is widely applied. Theoretical potential quantifies naturally available energy, while technical potential refines this based on technological extraction. Site-specific assessments employ high-resolution modeling combined with observational data (wave height, tidal velocity, etc.) to project energy yields. For instance, point absorber wave energy converters are evaluated using power matrices correlated with hourly wave data (significant wave height, peak period) to compute capacity factors. Performance indices like the **excess power index (IPeD)** in water networks identify technical inefficiencies, such as pressure-induced energy losses, enabling targeted optimization (Morani et al., 2024). Complementarity indices also quantify stability gains when integrating marine sources; for example, hybrid wind-wave-tidal systems on islands boost aggregate availability to > 70% by offsetting seasonal intermittency. Large-scale projects, such as China's Three Gorges Dam and Brazil's Itaipu Dam, dominate global capacity but exact heavy ecological and social tolls, including ecosystem fragmentation, displacement of communities, and sedimentation issues (Bin, 2023). In contrast, Norway and Switzerland emphasize decentralized, low-impact small hydropower systems, which minimize environmental disruption while supporting local energy resilience. These nations exemplify how technological choices reflect prioritization of ecological preservation over sheer output, though such systems often struggle to meet the escalating energy demands of larger economies. According to the International Energy Agency (IEA), the theoretical global potential of marine energy exceeds **29,500 TWh/year**, equivalent to 120% of global electricity demand in 2022 (International Energy Agency (IEA), 2023). However, technical and economic constraints currently limit its exploitation to coastal regions with high energy density, such as the UK, France, Canada, and South Korea. The technical potential of tidal stream energy alone is estimated at **1,200 TWh/year**, with wave energy contributing an additional **8,000–80,000 TWh/year** depending on technological advancements (International Renewable Energy Agency (IRENA), 2022). Regions like the European Atlantic coast, the Korean Strait, and the Bay of Fundy (Canada) are hotspots due to strong tidal currents. Meanwhile, OTEC's viability is greatest in tropical zones (Hawaii, Indonesia), where temperature gradients between surface and deep seawater exceed 20 °C.

Economic drivers similarly diverge between developing and developed nations. Economic viability is predominantly gauged through levelized cost of electricity (LCOE) projections, incorporating capital (CAPEX) and operational (OPEX) expenditures against capacity factors. Wave energy, currently high-cost, is forecast to reach < 70 €/MWh by 2035 in resource-rich regions, rivaling offshore wind (Bhuiyan et al., 2022). Techno-economic models integrate resource data, device efficiency, and cost-learning curves to simulate future competitiveness. System-level indices, such as global power balance metric, evaluate cost-benefits of interventions like pressure-reducing valves (PRVs) in hydropower networks. These quantify reductions in excess power dissipation (lowering IPeD from 61.7% to 17.5%), translating to energy savings and accelerated ROI (Jiang et al., 2023). Complementarity assessments further inform economics; co-locating wave energy with solar PV mitigates grid-balancing costs, enhancing project bankability. Multi-criteria methodologies also weigh environmental and regulatory constraints to derive practical potentials.

For coastal countries in South Asia and Africa, hydropower is a strategic tool for energy independence and economic growth, enabling electricity exports to neighboring regions. Laos, dubbed the "Battery of Southeast Asia," leverages the Mekong River's potential to export hydropower to Thailand and Vietnam, though this has sparked transboundary tensions over water governance (Vaidya et al., 2021). Meanwhile, developed countries such as the US and Canada (Helseth et al.,

2023) focus on retrofitting aging infrastructure, such as the Hoover Dam and Churchill Falls, to enhance efficiency and extend operational lifespans amid competing renewable investments). These retrofits underscore a shift from expansion to optimization in mature hydropower markets.

Climate change introduces further complexity, with impacts varying starkly across regions. Rising temperatures may temporarily boost Russia's hydropower output by 12% due to increased glacial meltwater, though this benefit is unsustainable as glaciers retreat. Conversely, drought-prone regions like Brazil face heightened volatility, as seen in the 2021 crisis at the Itaipu Dam, where output plummeted by 20%, forcing reliance on fossil fuels (Gonzalez-Salazar & Poganietz, 2022). Such climatic unpredictability challenges the reliability of hydropower as a baseload resource, necessitating adaptive strategies like diversified energy mixes and enhanced storage solutions.

Future projects highlight both ambition and risk. The Grand Inga Dam in the DRC, with a potential 39 GW capacity, could revolutionize Africa's energy landscape but requires unprecedented international cooperation and $80 billion in financing. Similarly, Russia's proposed Lower Tunguska project in Siberia aims to harness permafrost-affected rivers for 12 GW of power, yet faces formidable engineering and environmental hurdles. These ventures underscore the delicate balance between energy ambition and ecological stewardship, particularly in regions where governance and infrastructure are fragile (Table 4.1).

Future Potential and Areal Innovations

The marine energy sector is poised for exponential growth, driven by climate commitments and technological breakthroughs as follow: floating offshore wind (GWEC, 2023); tidal energy, wave energy, emerging technologies such as blue energy (salinity gradient) and marine biomass (algae-based biofuels) further expand the sector's scope. The IEA's *Net Zero by 2050* roadmap emphasizes marine energy as critical for decarbonizing island nations and remote coastal communities (*Climate Change and Energy Transition Law—Policies—IEA*, 2025) These projects collectively contribute < 1% to global electricity but are

Table 4.1 Hydropower Capacity by Region and Subregion (MW) project with 75 mv more power

Region	Subregion	Operating	Construction	Pre-construction	Announced	Prospective (Sum of Construction, Pre-construction, Announced)	Shelved	Mothballed	Retired	Cancelled
Global total		1,133,406	250,780	528,842	309,113	1,088,735	49,470	9,945	773	64,115
Africa	Northern Africa	6,123	2,735	0	1,000	3,735	0	0	0	0
	Sub-Saharan Africa	31,287	12,875	15,201	19,677	47,753	5,284	0	0	7,928
Americas	Latin America and the Caribbean	177,775	9,071	13,552	42,638	65,261	7,935	1,210	140	14,470
	Northern America	162,758	1,100	6,898	58,838	66,836	5,240	0	351	11,370
Asia	Central Asia	15,257	2,336	2,528	5,663	10,527	200	0	0	0
	Eastern Asia	372,686	172,161	309,852	81,558	563,571	5,862	82	0	1,339
	South-eastern Asia	44,573	11,869	38,352	20,015	70,236	4,152	6,077	0	9,847
	Southern Asia	70,164	31,963	112,971	57,255	202,189	11,680	2,071	0	9,926
	Western Asia	31,466	1,351	3,244	2,436	7,031	447	170	0	3,059
Europe	Eastern Europe	69,021	1,590	4,402	5,265	11,257	2,662	335	0	850

Region	Subregion	Operating	Construction	Pre-construction	Announced	Prospective (Sum of Construction, Pre-construction, Announced)	Shelved	Mothballed	Retired	Cancelled
	Northern Europe	44,801	0	4,486	4,710	9,196	0	0	0	600
	Southern Europe	45,390	519	8,471	867	9,857	2,068	0	0	2,151
	Western Europe	49,943	960	2,677	1,178	4,815	3,615	0	282	550
Oceania	Australia and New Zealand	12,005	2,250	5,348	7,935	15,533	325	0	0	225
	Melanesia	157	0	860	78	938	0	0	0	1,800
	Micronesia	0	0	0	0	0	0	0	0	0
	Polynesia	0	0	0	0	0	0	0	0	0

Source Summary Tables, Global Energy Monitor (2025)

expected to grow to 3–5% by 2040 (International Energy Agency (IEA), 2023), balances ecological preservation with economic growth. Strategic investments, cross-border collaboration (EU-ASEAN ocean energy partnerships), and community engagement are essential to unlock its full potential (Agyekum et al., 2024) (Table 4.2).

Key Vulnerabilities in the Water and Energy Sector

See Table 4.3.

Barriers to MRE Implementation

Accurate evaluation of MRE projects remains challenging due to the heterogeneous economic and market challenges with holistic lifecycle assessments must account for intersectoral trade-offs especially in case of using hybrid models with wave energy converters (WECs) may enhance energy security but disrupt aquaculture and fisheries, which contribute $362 billion annually to the global economy. In Scotland, stakeholder consultations revealed conflicts between MRE developers and fishing communities over seabed access, necessitating participatory spatial planning frameworks. Environmental risks, such as electromagnetic field (EMF) emissions from subsea cables and noise pollution during installation, further complicate regulatory approvals.

MRE technologies remain cost-prohibitive in developing nations, where 90% of projects are in pre-commercial stages. Cross-country analyses indicate that MRE adoption correlates strongly with national policy incentives (feed-in tariffs, tax rebates) (Apolonia et al., 2021). For example, South Korea's *Tidal Power Plant Project* reduced LCOE by 40% through state-backed R&D partnerships, while Chile's lack of targeted subsidies has stalled OTEC development. Workforce training gaps also persist; a 2022 IRENA report noted that < 15% of MRE firms in Southeast Asia have access to specialized technicians.

Table 4.2 Marine and offshore energy potential by country

Country	Energy type	Technical potential (TWh/year)	Economic feasibility	Notable projects	Key challenges
United Kingdom	Tidal Stream	34	High	MeyGen (398 MW), Perpetuus Tidal Energy Centre	High CAPEX, environmental monitoring
	Wave Energy	50–75	Moderate	WaveHub, EMEC Orkney	Technology immaturity, survivability in storms
	Floating Offshore Wind	2,300	High	Hywind Scotland (30 MW), Dogger Bank (3.6 GW)	Grid integration, supply chain bottlenecks
France	Tidal Energy	26	Moderate	Raz Blanchard Tidal Farm (17 MW), NEMO OTEC	Environmental pushback, sediment dynamics
Canada	Tidal Stream	35	High	FORCE Bay of Fundy, Tocardo Tidal Turbines	Harsh marine conditions, seasonal variability
South Korea	Tidal Barrage	520	Moderate	Sihwa Lake (254 MW), Uldolmok Tidal Plant	Ecosystem disruption, siltation

(continued)

Table 4.2 (continued)

Country	Energy type	Technical potential (TWh/year)	Economic feasibility	Notable projects	Key challenges
	Offshore Wind	1,200	High	Donghae 1 (200 MW), Jeju Island	Space competition with fisheries
USA	Wave Energy	2,640	Moderate	PacWave (Oregon), Hawaii OTEC Pilot	Regulatory delays, high grid connection costs
	Floating Offshore Wind	4,200	High	Vineyard Wind (800 MW), California Floating	Deepwater engineering challenges
Indonesia	Ocean Thermal (OTEC)	240	Low-Moderate	Bali OTEC Pilot Plant	High CAPEX, lack of localized supply chains
Australia	Wave Energy	1,300	Moderate	CETO 6 (Carnegie Clean Energy)	Remote locations, transmission infrastructure gaps
Norway	Salinity Gradient	200	Low	Statkraft Osmotic Power Prototype	Technology immaturity, low energy density

Country	Energy type	Technical potential (TWh/year)	Economic feasibility	Notable projects	Key challenges
China	Tidal & Offshore Wind	1,300	High	Zhoushan Tidal-Wind Hybrid, Yangshan Wind Farm	Environmental impact, marine traffic conflicts
Japan	Floating Offshore Wind	30	Moderate	Fukushima Forward (7 MW), Kairyu Tidal	Seismic risks, typhoon resilience

Sources International Renewable Energy Agency (IRENA) (2022), *Marine Energy Capacity Worldwide, 2024* (2024)

Table 4.3 Ecosystem impacts across lifecycle stages for hydropower and marine energy

Lifecycle stage	Hydropower impacts	Marine energy impacts	Comparative analysis
1. Planning/Installation	- Habitat fragmentation from reservoir inundation - GHG emissions (CH_4 from organic decay)	- Seabed disturbance from anchoring - Noise pollution displacing marine mammals	Hydropower impacts are irreversible (for example, submerged forests), MES impacts are localized but recoverable
2. Construction/Deployment	- Sediment trapping alters downstream geomorphology - Displacement of riparian species	- Artificial reef effects boost biodiversity - Electromagnetic field (EMF) emissions	Both disrupt benthic systems, but MES offers compensatory habitat benefits
3. Operation	- Hydropeaking disrupts fish migration - Thermal stratification reduces oxygen	- Collision risks for marine fauna - Altered currents affect plankton distribution	Hydropower impacts are systemic (entire watersheds); MES impacts are micro-scale
4. Decommissioning/End-of-Life	- Sediment release revitalizes habitats - Short-term benthic smothering	- Habitat loss from infrastructure removal - Legacy pollution risks	Hydropower decommissioning often benefits ecosystems; MES removal erases artificial reefs
5. Cumulative Impacts	- Synergy with climate change (warming reservoir hypoxia)	- Ocean acidification exacerbates corrosion and habitat suitability	Both face amplified risks from climate stressors, requiring adaptive design

Sources Gemechu and Kumar (2022), Grill et al. (2015), *Hydropower Sustainability Assessment Protocol (2025)*, *Offshore Wind Energy Impacts on Fishing Communities | NOAA Fisheries* (2025), IPCC (2023)

Evaluation of marine renewable energy (MRE) projects remains challenging due to the heterogeneous nature of marine energy systems. Capital expenditures (CAPEX) and operational expenditures (OPEX) are often miscalculated because MRE installations integrate diverse subsystems, such as mooring systems, energy converters, and grid infrastructure, each with distinct cost structures (Allan et al., 2014). The levelized cost of energy (LCOE), the primary metric for assessing economic feasibility, currently ranges between 150–300/MWh for wave energy and 130–280/MWh for tidal stream systems significantly higher than offshore wind (80–120/MWh) and solar P V (80–120/MWh) and solar PV (30–60/MWh). This cost disparity stems from low technology readiness levels (TRL 4–6 for most MRE systems) and the absence of standardized deployment protocols. Pilot-scale projects dominate the sector, with fewer than 10% of initiatives reaching commercial scale, hindering economies of learning and scale.

Environmental Vulnerabilities

Marine energy infrastructure is inherently exposed to dynamic oceanic and climatic systems. **Climate change-induced hazards**, such as rising sea levels, intensified storm surges, and ocean acidification, threaten the structural integrity of offshore installations. For instance, tropical cyclones in the South China Sea and North Atlantic have damaged floating wind turbines and tidal arrays, highlighting the sector's exposure to extreme weather (IPCC, 2023). While MRE is lauded for low GHG emissions, its ecological impacts are under-theorized. Rockström's (Brockhaus et al., 2019) framework warns that marine energy extraction could destabilize oceanic biogeochemical cycles. Tidal barrages, such as South Korea's Sihwa Lake plant, disrupt sediment transport, exacerbating coastal erosion. Additionally, biofouling on devices like the CETO wave energy system introduces invasive species risks Ecological risks include habitat fragmentation from tidal barrages, noise pollution affecting marine mammals, and sediment displacement altering coastal ecosystems.

Technical and Operational Risks

Technological stagnation in MRE stems from lock-in effects favoring incremental improvements over radical innovation. Entrenched systems as an offshore oil infrastructure resist disruption, as seen in the dominance of horizontal-axis tidal turbines despite superior efficiency in vertical-axis designs. Bibliometric analyses reveal a disproportionate focus on tidal energy, neglecting wave and OTEC systems (Agyekum et al., 2024). Marine energy technologies face **high failure rates** due to harsh operating conditions. Corrosion from saline environments, biofouling, and mechanical stress on wave energy converters (WECs) reduce efficiency and lifespan. For example, the collapse of the *Pelamis Wave Energy Converter* in Portugal underscored the challenges of scaling prototype technologies to commercial viability (Ocean Energy Systems (OES), 2023).

Technical Innovations and Limitations OF MRE and Hydro Energy

Blue economy integrated with circularity and ecosystems services: tidal tech merge energy production with marine conservation. Marine and hydro energy technologies are undergoing rapid transformation, driven by regional resource availability and regional diversity rather than uniform structures. Future common global innovations includes AI and hybrid Systems with machine learning which optimizes marine energy forecasting. For following circular production principles used advanced recyclable materials and 3D-printed turbines (Finnrunner's composite designs) cut costs by 20% (Ghandehariun et al., 2024) (Table 4.4).

Table 4.4 Modern innovations in marine and hydro renewable energy

Technology	Description	Key features	Continental suitability
Fish-Safe Hydropower Turbines	Turbines with rounded blades allowing >99% fish survival, mimicking natural river conditions	Enhances biodiversity; retrofittable to existing plants	North America, Europe
Autonomous Offshore Power Systems (AOPS)	Wave energy converters (e.g., C-Power's SeaRAY/StingRAY) for low/high-power offshore needs	Supports subsea assets; integrates with blue economy sectors	North America, Europe
Submerged Wave Energy Converters	Fully submerged systems (e.g., CalWave × 1™) surviving extreme conditions without visual impact	Storm-resilient; modular architecture	North America, Asia–Pacific
Modular Riverine Hydrokinetic Systems	ORPC's RivGen® units for remote riverine communities, replacing diesel generators	Scalable; minimal ecological disruption	South America, Africa
Tidal Energy TriFrame™ Turbines	Multi-turbine systems (e.g., Verdant Power) achieving >46% water-to-wire efficiency	High reliability; grid-compatible	North America, Europe
Hybrid DG Systems Optimization	Models integrating solar, wind, and storage for off-grid applications	Reduces fossil dependency; supports SDGs	Africa, Asia

(continued)

Table 4.4 (continued)

Technology	Description	Key features	Continental suitability
G-res GHG Emissions Tool	Calculates net emissions from reservoirs, aligning with green bond criteria	Cost-effective; supports climate finance	Global (e.g., Brazil, Canada)
Marine Energy Atlas	DOE's portal for resource assessment (wave, tidal, thermal)	Data-driven site selection; quantifies technical potential	North America, Europe
Solid Oxide Fuel Cell Hybrids	Combines fuel cells with renewables for efficient baseload power	High efficiency; reduces grid instability	Asia, Europe
Pumped Storage Hydropower (PSH)	Energy storage using water reservoirs, providing 24 GW of U.S. grid flexibility	Long-duration storage; stabilizes renewable grids	Globally (U.S., China)
Salinity Gradient Energy	Harnesses osmotic power from seawater–freshwater interfaces	Untapped potential; suitable for estuaries	Europe, South America
Ocean Thermal Energy Conversion (OTEC)	Exploits temperature gradients for power, with U.S. potential equivalent to 98% of 2019 electricity	Baseload capability; tropical regions	Asia–Pacific, Caribbean
Vehicle-to-Grid (V2G) Integration	EV batteries as grid storage in hybrid microgrids	Balances demand peaks; reduces infrastructure costs	Europe, North America
AI-Driven Predictive Maintenance	Machine learning models for hybrid system optimization	Minimizes downtime; extends asset lifespan	Global
Community-Based Microgrids	Solar/wind hybrids with battery storage for rural electrification	Enhances energy access; aligns with local governance	Africa, South Asia

Sources Thorson et al. (2022), Phanindra et al. (2024), Wheatley (2024), Open Access Government (2024), Multon (2012), Mazi et al. (2021), Seriño (2021), Nugraha and Priyambodo (2021), Bhatt and Dhyani (2015)

Cross Cultural Observation

North America: Balancing Ecological and Technological Ambitions

The US exemplifies the tension between scaling marine energy and ecological preservation. Regions like Alaska and the West Coast possess tidal and wave energy potentials exceeding 2,300 TWh/year, yet projects such as PacWave (Oregon) and Bourne Tidal Test Site (Massachusetts) prioritize minimizing ecological disruption (Marine Energy Conference *2025* | *Marine Energy Conference 2025* | *Marine Energy Events 2025* | *Marine Energy Meetings 2025*, n.d.). Fish-safe turbines (e.g., Natel Energy's rounded-blade designs) achieve > 98% fish survival rates while retrofitting aging hydropower infrastructure, boosting output by 10%. The U.S. faces environmental regulations challenges complicate rapid infrastructure deployment 10. However, permitting delays, exemplified by federal pauses on wind projects, threaten progress despite state-level commitments like California's 100% clean energy target by 2045. In North America region used autonomous offshore power systems (AOPS): C-Power's SeaRAY (*Ocean Power Products—SeaRAY & StingRAY AOPS* | *C-Power*, 2025) integrates wave energy with blue economy sectors (subsea data centers) and ocean thermal energy conversion (OTEC): U.S. projects target tropical zones, leveraging temperature gradients for baseload power. In this case, seasonal variability in ocean currents necessitates hybrid systems combining storage (pumped hydro) with AI-driven predictive maintenance.

The PacWave South test site, developed with input from Yurok Tribal leaders, includes adaptive management clauses to halt operations if ecological thresholds are breached. This participatory model has become a blueprint for U.S. offshore projects and demonstrates resource variability (5% of global tidal sites exceed 2.5 m/s flow velocities required for commercial tidal turbines); technological maturity with TRL lags 10–15 years behind offshore wind also shows the issues with environmental durability: saltwater corrosion, biofouling, and extreme weather events increase maintenance costs by 25–40% compared to terrestrial renewables (*South Test Site—PacWave*, 2025).

EU-Wide Initiatives

Horizon 2020 and REPower EU The Horizon 2020 program has allocated €30 billion for research and innovation realized locally, with a focus on sustainable energy solutions (*Horizon 2020-Funded EPIC Project Releases Policy Recommendations for Europe and New Zealand's AI Collaboration Future | News | CORDIS | European Commission*, 2025). The EU's Green Deal prioritizes marine energy, with Scotland's tidal projects and Mediterranean OTEC research leading innovation. Strict environmental mandates, such as the Water Framework Directive (*Water Framework Directive—European Commission*, 2025), have spurred fish-friendly retrofits and salinity gradient systems in estuaries like the Rhine-Meuse Delta. Pumped storage hydropower (PSH) dominates grid flexibility, contributing 90% of Europe's energy storage capacity. A dedicated €40 million fund supports tidal energy projects, including TIGER (Tidal Stream Industry Energiser Project) (*TIGER: Tidal Stream Industry Energiser Project*, 2025) in the English Channel, which aims to reduce costs of tidal stream energy by 40% by 2030 emerging Vehicle-to-Grid (V2G) integration and digital twins. The basic barriers in Europe are an aging hydropower infrastructure requires €50 billion in upgrades to meet 2050 targets and public opposition to large-scale projects (offshore wind farms) persists due to cultural and economic concerns. But locally realized many systemic initiatives.

The REPowerEU initiative further accelerates MRE adoption to reduce reliance on Russian fossil fuels, targeting 10 GW of ocean energy capacity by 2030. OceanDEMO (Demonstration Programme for Ocean Energy) tests hybrid wave-tidal systems in Western Europe, while OESA (Ocean Energy Scale-Up Alliance) focuses on large-scale deployment in the North Sea.BlueGIFT (Blue Growth and Innovation Fast Tracked) in the Atlantic promotes open-sea testing of emerging technologies like floating wind turbines and oscillating water columns. In France NH1 Tidal Energy Project (*Normandie Hydroli-ennes—Projet NH1*, 2025): Funded by the EU Innovation Fund (€31.3 million), this project deploys four AR3000 tidal turbines (3 MW each) in Normandy, generating 34 GWh/year for 15,000 households. The turbines, designed by Proteus Marine Renewables, are 80% domestically produced and minimize ecological impact. Experimental Contracts for Renewables: Simplified permitting and "green tariffs" for small-scale MRE projects, aligning with France's goal to achieve 5 GW of tidal energy by 2030. Ireland developing ocean Renewable Energy Development Plan (OREDP) (*Offshore Renewable Energy Development Plan (OREDP)*, 2025): Integrates MRE into national carbon budgets, emphasizing wave energy for Marine Energy Test Site (AMETS) trials technologies like WaveRoller (oscillating wave surge converters) and SEAI-funded tidal arrays in Galway Bay (Blain, 2024). Italy Blue Technology Cluster combines shipbuilding expertise with MRE innovation (*Cluster collaboration*, 2025). Projects include Inertial Sea Wave Energy Converters (ISWEC) in the Mediterranean, which use gyroscopic systems to capture wave energy (Khedkar et al., 2021).

Belgium focuses on sustainable blue economy partnerships, including floating offshore wind farms in the North Sea and seabed-anchored tidal systems. Sweden developing National Ocean Energy Program (2018–2024) (Ocean Energy Systems (OES), 2023): Allocated €10.4 million to 21 projects, such as CorPower Ocean's Wave Energy Converters, which mimic heartbeats to optimize energy capture in harsh Baltic conditions (*CorPower Ocean—Wave Power. To Power the Planet.*, 2021). In Spain the PLOCAN test site (*South Test Site—PacWave*, 2025) in the Canary Islands hosts PTO (Power Take-Off) systems for hybrid wave-wind energy. Portugal oriented on free technological zones: Offshore test sites like WindFloat Atlantic (25 MW floating wind farm) and Waveroller installations near Peniche demonstrate grid-connected wave energy solutions (*Windfloat Atlantic | Offshore Wind Energy*, 2025). The Floating Wind-Wave Hybrid Project in Portugal achieved a 22% capacity factor increase by co-locating WECs with floating turbines. Patent filings for MRE technologies surged to 200+/year post-2020, focusing on modular designs and AI-driven predictive maintenance. North part of Europe included Wave Energy Scotland

(WES) (Wave energy Scotland, 2025): Funds R&D for Pelamis Wave Energy Converters and MeyGen Tidal Array (*MeyGen Tidal Energy Project | Tethys*, n.d.), the world's largest tidal stream project (398 MW planned). Orkney's European Marine Energy Centre (EMEC) (*EMEC: European Marine Energy Centre*, 2025) has transformed the region into a global testbed for tidal and wave energy. Local schools collaborate with EMEC on STEM programs, while revenue from leasing seabed rights funds community infrastructure. The project exemplifies economic growth with marine conservation. The Wales (UK) secured long-term funding for Tidal Lagoon Swansea Bay (320 MW) (*Swansea Bay Tidal Lagoon (SBTL) | Tethys*, 2025) and Morlais Demonstration Zone, testing submerged tidal kites.

Asia–Pacific: Tropical Innovation and Hybrid Systems

Technical challenges in South Asia with deep-water tidal sites (Japan's Kuroshio Current) demand advanced mooring systems, raising costs by 30% and regulatory fragmentation slows cross-border energy trading in ASEAN nations. Tropical countries like Indonesia and the Philippines leverage OTEC for baseload power, while Japan pioneers floating solar to address land scarcity. Solid oxide fuel cells and AI-driven microgrids enhance energy access in remote island communities, reducing reliance on diesel by 40% in the Philippines. Singapore's REIDS initiative (*Renewable Energy Integration Demonstrator—Singapore Flagship Programme*, n.d.) serves as a regional hub for sustainable energy solutions in Southeast Asia. Managed by Nanyang Technological University, it focuses on hybrid microgrid systems integrating solar, wind, tidal energy, and power-to-gas technologies. Key projects include low voltage microgrid cluster (LVMGC) with investigatiom of grid resilience and energy trading in tropical climates; grid-of-grids architecture: tests dynamic system optimization and cybersecurity for off-grid communities, supporting decarbonization in archipelagic regions. South Korea aims to deploy 1.2 GW of ocean energy by 2030 through offshore test Sites: open-sea platforms for wave and tidal energy prototypes. South Korea's Sihwa generating 552 GWh/year (*Technology Case Study*, 2025), improved water quality by flushing polluted sediments, boosting fish stocks and tourism. This aligns with Sen's human development approach, as improved health and livelihoods followed energy access. China's initiatives focus on Solar-wave-diesel microgrids in the South China Sea, realizing Zhoushan Tidal Plant (3.4 MW) and wave energy converters in Guangdong and India: Deep Ocean Mission (DOM) (*Deep Ocean Mission | India Science, Technology & Innovation—ISTI Portal*, 2025) with Subsea turbines for low-velocity currents and offshore desalination plants, allocated for deep-sea exploration and energy harvesting.

Africa: Decentralized Solutions and Resource Potential

The obstacles in Africa region mostly deals with economical underfunded off-grid solutions attract only 3% of continental investments and domination of the fossil fuels (69% of 2025 capacity), necessitating blended finance models for green hydrogen megaprojects. Despite 600 million lacking electricity, Africa's renewable capacity grew to 30 GW by 2024, driven by hydropower in Zambia and solar/wind in South Africa. Modular hydrokinetic systems (ORPC's RivGen®) ("RivGen® Power System & Integrated Microgrid Solutions," 2025) electrify Congo River communities, displacing diesel generators (*Accelerating Renewable Energy Deployment with Regional Integration*, 2025).

Now pioneering projects including Kudu Gas-to-Power (Namibia) (*Ministry of Industries, Mines and Energy—Gas to Power Project*, 2025): Hybridizes gas with renewables for grid stability and OceanHub Africa (*Home—OceanHub Africa*, 2025): Accelerates startups like Kinetic NRG (*Home—KineticNRG | Green Energy Supplier*, 2025), which develops low-velocity hydrokinetic kits.

South America: Hydropower Modernization and Ecological Balance

South America potential in hydro and marine renewable energy meets with the risk of deforestation and indigenous land conflicts hinder project approvals. Despite of that, Brazil and Chile use hybrid Micro-grids with solar-hydro-storage systems (in Chile's Atacama Desert) and Brazil microhydro statiom enhance reliability (Tafur, 2011). Also in use dynamic pumped hydro: RheEnergise's (*RheEnergise High-Density Hydro—How It Works*, 2025) high-density fluid enables low-elevation storage in the Andes. Climate-induced droughts reduce hydropower output by 15% in Brazil 11. Colombia (70% hydropower) and Brazil (300 MW added in 2022) exemplify efforts to modernize aging dams while mitigating ecological impacts. The G-res tool quantifies reservoir emissions, aligning projects with green bond criteria (*G-Res* Tool, 2025). Amazonian rivers offer 78 TWh/year in hydrokinetic potential, yet sediment transport disruptions threaten aquatic ecosystems.

Contemporary Water Management Principles

1. **Definition of Energy Management for Hydropower and Marine Energy**

Energy Management for Hydropower and Marine Energy refers to the integrated set of *technical, operational, and economic processes* applied throughout the lifecycle of facilities harnessing energy from water flows (rivers, reservoirs, tides, waves, ocean currents). Its core objective is to optimize the safe, reliable, efficient, and sustainable conversion of hydraulic and kinetic hydro energy into electricity, while balancing resource availability, environmental constraints, grid requirements, and economic viability (International Renewable Energy Agency (IRENA), 2022; Ocean Energy Systems (OES), 2023). Effective marine and hydro energy governance necessitates a systemic integration of four core principles. First, **adaptive management**, grounded in resilience approach, employs flexible, iterative planning to address ecological and socio-technical uncertainties, utilizing dynamic environmental monitoring (AI-powered sensors for real-time biodiversity tracking (Frid et al., 2011) and modular, removable designs (like floating OTEC platforms) to minimize ecological footprints. Second, **community-based participatory approaches** actively incorporate local knowledge and cultural values, reducing socio-economic risks, which engaged Indigenous leaders in co-developing wave energy aligned with traditional stewardship. Third, **transboundary collaboration** and **policy integration** are facilitated through regional partnerships like the North Seas Energy Cooperation (*The North Seas Energy Cooperation*, 2025) (NSEC), harmonizing offshore grids and risk-sharing across the EU, UK, and Norway, while institutions like IRENA advance marine spatial planning tools to resolve conflicts between ocean uses (shipping vs. energy zones). Finally, financial de-risking mechanisms, mobilizing projects in for realization (*Financing Climate Futures | OECD*, 2025).

2. **Goals of Energy Management for Hydropower and Marine Energy**

Based on the core principles of management and adapted to sector specifics, the primary goals are:

1. **Maximize Energy Yield and Revenue:** Optimize power generation considering hydrological variability (rainfall, snowmelt), tidal cycles, wave climate forecasts, and electricity market dynamics (spot prices, ancillary services).
2. **Ensure System Reliability, Safety and Resilience:** Maintain structural integrity of dams, penstocks, tidal barrages, wave energy converters (WECs), and tidal turbines; ensure operational safety in harsh marine environments; implement robust monitoring and emergency protocols (*IEC TS 62600-2*, 2025), (*2023 World Hydropower Outlook*, 2025)
3. **Optimize Operational and Maintenance Costs:** Develop predictive maintenance strategies to minimize downtime and repair costs, particularly critical for offshore marine energy devices with high access costs and challenging logistics.
4. **Facilitate Grid Integration & Stability:** Manage variable and sometimes predictable (tides) output; provide grid services (frequency regulation, inertia—especially from hydropower); ensure power quality compliance.

5. **Manage Environmental & Social Impacts:** Operate within ecological flow requirements (hydropower); minimize impacts on marine ecosystems (noise, habitat, marine mammals); engage with local communities; manage sedimentation (hydropower) (*IPCC Working Group I Summary for Policymakers in the UN Official Languages— IPCC*, 2025).

6. **Optimize Water Resource Utilization (Hydropower):** Balance energy production with irrigation, flood control, drinking water supply, and recreational needs through sophisticated reservoir management (*2023 World Hydropower Outlook*, 2025).

7. **Demonstrate Technological Viability & Performance (Marine Energy):** Collect and analyze operational data to validate performance, reduce LCOE, and inform future design iterations, crucial for this emerging sector.

3. Methods and Tools

Leveraging core RE management methods with domain-specific adaptations:

- **Resource Assessment & Forecasting:** Advanced hydrological modeling, tidal harmonic analysis, numerical wave modeling (SWAN, MIKE) (Wiloso & Heijungs, 2013), real-time meteorological/oceanographic (metocean) data buoys.

- **Technological Monitoring and Diagnostics:** SCADA systems; vibration analysis for turbines; strain gauges on WECs; corrosion monitoring; underwater inspections (ROV/AUV); power quality monitoring (*Supervisory Control and Data Acquisition System—an Overview | ScienceDirect Topics*, 2025).

- **Predictive and Condition-Based Maintenance:** Utilizing sensor data (vibration, temperature, lubricant analysis) and AI/ML algorithms to predict failures, especially vital for hard-to-access marine devices.

- **Performance Analysis and Optimization:** Calculation of Key Performance Indicators (KPIs): Plant Availability, Capacity Factor

(CUF), Performance Ratio (PR—for marine), Water-to-Wire efficiency (hydropower). Hydro: Unit commitment optimization, reservoir optimization models. Marine: Power-take-off (PTO) system control optimization (Frid et al., 2011).

- **Operational Planning and Scheduling:** Hydro: Day-ahead/week-ahead scheduling integrating market prices and water inflow forecasts. Marine: Scheduling maintenance windows based on weather/sea-state forecasts (wave height, period).
- **Asset Lifecycle Management:** Risk-based inspection (RBI) planning, life extension strategies for aging hydropower assets, lifecycle cost modeling for marine energy (Ocean Energy Systems (OES), 2023)
- **Environmental Compliance Monitoring:** Real-time water quality sensors (dissolved oxygen, temperature), fish passage monitoring, marine mammal acoustic monitoring (PAM), sediment transport modelling (*Special Report on the Ocean and Cryosphere in a Changing Climate*, 2025).
- **Smart Water Systems—IoT and AI-Driven Networks**, decentralized systems with modular wastewater aligns with circular economy principles.
- **Implementation of the nature-based solutions (NBS):** reflecting sociohydrology.
- **Desalination Innovations**—graphene oxide membranes, solar-powered desalination
- **Hydropower Modernization**—small modular hydropower, micro-turbines (<10 MW) electrify remote regions, pumped storage hydropower, fish-friendly turbines
- **Marine Energy** witn oscillating water columns, tidal stream farms, floating solar-wave
- **Integrated Water-Energy-Food** (WEF) Nexus measurement, polycentric governance (An, 2023)
- **Climate-Resilient Infrastructure** with dynamic adaptive pathways and hybrid renewable systems (*Special Report on the Ocean and Cryosphere in a Changing Climate*, 2025).
- **Blue Energy Innovations**—salinity gradient power, hydrogen from hydropower

4. Level of Impact and Key Actors

- **Level of Impact:** Primarily **Project, Facility, and Corporate/Portfolio Level.** Decisions focus on the specific power plant, array of tidal turbines, wave farm, or a utility's portfolio of hydro/marine assets.
- **Key Executing Actors:**
  - **Plant/Facility Managers an Operators:** Day-to-day control room operations, local monitoring.
  - **Engineers:** Civil (dam, structure), Mechanical (turbines, PTO), Electrical (generators, grid connection), Environmental.
  - **Technicians:** Field maintenance, inspection crews (including specialized marine technicians/divers).
  - **Asset Managers:** Strategic oversight of portfolios, investment decisions, risk management, lifecycle planning.
  - **O&M Contractors:** Specialized service providers, particularly prevalent in marine energy.
  - **Energy Traders/Market Operators:** (Especially for flexible hydropower)—bidding generation into markets.
  - **Environmental Specialists:** Monitoring compliance, implementing mitigation measures.
  - **Technology Developers (Marine Energy):** Closely involved in early-project management for performance validation.

5. Social Impact and Viability Energy Management

MRE's and hydro energy economic viability is hamstrung by path dependencies favoring fossil fuels and terrestrial renewables. MRE technologies remain cost-prohibitive in developing nations, where 90% of projects are in pre-commercial stages. Cross-country analyses indicate that MRE adoption correlates strongly with national policy incentives (e.g., feed-in tariffs, tax rebates) and R&D investments. For example, South Korea's Tidal Power Plant Project reduced LCOE by 40% through state-backed R&D partnerships, while Chile's lack of targeted subsidies has stalled OTEC development. Workforce training gaps also persist; a 2022 IRENA report noted that < 15% of MRE firms in Southeast Asia have access to specialized technicians. But the economic benefits includes job creation (The UK's European Marine Energy Centre (EMEC) (*EMEC: European Marine Energy Centre*, n.d.) has generated 1,200+ jobs in Orkney, with 30% in high-skilled engineering roles). Marine and hydro energy projects generate employment across the value chain, from manufacturing and installation to maintenance and research. The global transition to renewable energy has catalyzed significant employment opportunities across marine, oceanic, and hydropower sectors. These industries are poised to play a pivotal role in achieving climate goals while fostering socio-economic development.

This analysis synthesizes current statistics, growth projections, and regional dynamics, drawing on recent reports from international agencies and research institutions. Hydropower, marine, and oceanic energy sectors collectively represent a cornerstone of the global green job market, with 6.5–8.5 million potential jobs by 2050. Realizing this potential demands coordinated policy frameworks, workforce training, and equitable resource allocation (IRENA, 2023). Regional collaboration and innovation in decentralized systems will be vital to ensuring inclusive growth and climate resilience. Hydropower remains the largest renewable energy employer, with 2.5 million direct jobs globally in 2022 (Global Maritime Forum, 2023), primarily in operations, maintenance, and construction. Asia dominates this sector (You Need These Skills to Get That Green Job, 2022). The net-zero target by 2050, could generate 4 million cumulative jobs by 2050, predominantly in renewable energy infrastructure and e-fuel production (e.g., green hydrogen and ammonia) Renewable energy infrastructure: building 2 TW of offshore wind and solar capacity by 2050 could create 1.5–3 million jobs in the 2030s, focusing on manufacturing (7,500 job-years/GW for solar PV) and construction (17,500 job-years/GW for solar PV) 10. Green hydrogen and ammonia supply chains may generate 100,000–1 million jobs by the 2030s, contingent on scaling electrolyzer capacity and fuel synthesis facilities (CSIS, 2023). Regional disparities persist, with Europe leading in offshore wind technology and Asia dominating manufacturing. However, Africa and Latin America are poised to benefit from untapped offshore wind and hydropower resources, provided they develop local supply chains. While hydropower employment is stable, marine energy jobs face future uncertainty due to automation and AI adoption. Fossil fuel-dependent regions, particularly in the Global South, risk job losses without retraining programs. The ILO estimates a net gain of 18 million jobs in renewables by 2030, offsetting declines in fossil sectors (United Nations, 2023).

Hybrid MRE-wind installations, such as Denmark's Kriegers Flak, attract eco-tourism, boosting local GDP by 8–12% in coastal regions. Stiglitz's (2002) elucidates how global financial architectures—such as IMF loan conditionality—disincentivize developing nations from investing in high-risk MRE projects. For example, Ghana's abandoned 150 MW wave energy initiative faltered due to austerity-driven budget cuts. Incumbents like offshore wind lobbyists capture subsidies, marginalizing MRE in policy discourse. The sector's reliance on bespoke, small-scale prototypes (oscillating water columns) inflates capital costs, perpetuating a cycle of underfunding (Shove & Walker, 2010). Marine energy projects often encounter resistance from coastal communities reliant on traditional livelihoods, such as fishing and tourism. In Scotland, the *Caitlín Ness* tidal project faced opposition from fishers concerned about gear entanglement and restricted access to fishing grounds. Similarly, Indigenous communities in Canada's Bay of Fundy have raised cultural sovereignty concerns over tidal energy development disrupting ancestral waters. Energy projects that integrate traditional knowledge foster cultural resilience. In Canada's Bay of Fundy, the FORCE tidal energy initiative partnered with the Mi'kmaq First Nation to monitor ecological impacts while reviving ancestral fishing practices. Economic disparities further exacerbate risks. Developing nations, such as Indonesia and Small Island Developing States (SIDS), lack the capital and infrastructure to adopt advanced marine technologies, perpetuating reliance on fossil fuels, and energy justice frameworks are critical to address these inequities.

The management examination of hydropower and marine energy systems underscores the interconnected relationship between technological innovation and social transformation. Structural analyses reveal persistent inequities in terms of marginalization of indigenous voices in tidal energy projects or the uneven distribution of marine technology financing which is rooted in colonial legacies and neoliberal governance. Conversely, grassroots mobilizations, participatory spatial planning frameworks, and hybrid governance models demonstrate the agency of communities in redefining energy justice. Historically, the evolution of these sectors reflects path dependencies shaped by era infrastructure investments and integrated climate policy shifts, illustrating how institutional memory both enables and constrains sustainability agendas. Ultimately, the trajectory of hydropower and marine energy hinges on transcending reductionist paradigms to integrate structural equity, participatory governance, and intergenerational ethics for aligning energy futures with the principles of planetary boundaries and social solidarity.

References

2023 World Hydropower Outlook. (2025). Retrieved June 2, 2025, from https://www.hydropower.org/publications/2023-world-hydropower-outlook.

Accelerating renewable energy deployment with regional integration: A Retrospective on Southern Africa's Energy Transition—RES4Africa. (2025, February 18). Retrieved from https://res4africa.org/news/2025/accelerating-renewable-energy-deployment-with-regional-integration-a-retrospective-on-southern-africas-energy-transition/.

Agyekum, E. B., Khan, T., Dankwa Ampah, J., Giri, N. C., Fendzi Mbasso, W., & Kamel, S. (2024). Review of the marine energy environment-a combination of traditional, bibliometric and PESTEL analysis. *Heliyon, 10*(6), e27771. https://doi.org/10.1016/j.heliyon.2024.e27771

Allan, G. J., Lecca, P., McGregor, P. G., & Swales, J. K. (2014). The economic impacts of marine energy developments: A case study from Scotland. *Marine Policy, 43*, 122–131. https://doi.org/10.1016/j.marpol.2013.05.003

An, D. (2023). A simultaneous equations approach to analyze the sustainable water–energy–food nexus in South Korea. *Environmental Research Communications, 5*(9), Article 095017. https://doi.org/10.1088/2515-7620/acfb61

Apolonia, M., Fofack-Garcia, R., Noble, D. R., Hodges, J., & Correia da Fonseca, F. X. (2021). Legal and political barriers and enablers to the deployment of marine renewable energy. *Energies, 14*(16). https://doi.org/10.3390/en14164896

Bhatt, K. K., & Dhyani, R. (2015). *Communication and its role in promoting micro hydro based green energy solutions.* Retrieved from https://api.semanticscholar.org/CorpusID:114859895.

Bhuiyan, M. A., Hu, P., Khare, V., Hamaguchi, Y., Thakur, B. K., & Rahman, M. K. (2022). Economic feasibility of marine renewable energy: Review. *Frontiers in Marine Science.* https://doi.org/10.3389/fmars.2022.988513

Bin, A. (2023). Assessing the role of sustainable construction practices in the one belt one road initiative: A comparative analysis of China and Southeast Asian countries. *Journal of Digitainability, Realism & Mastery (DREAM).* Retrieved from https://api.semanticscholar.org/CorpusID:258814470.

Blain, L. (2024, February 2). *WaveRoller sea-floor generator approaches commercial deployment.* New Atlas. Retrieved from https://newatlas.com/energy/waveroller-wave-energy/.

Brockhaus, S., Petersen, M., & Knemeyer, A. M. (2019). The fallacy of "trickle-down" product sustainability. *International Journal of Operations & Production Management, 39,* 1166–1190. https://doi.org/10.1108/IJOPM-03-2019-0181

Climate change and energy transition law—Policies—IEA. (2024). Retrieved March 1, 2024, from https://www.iea.org/policies/13323-climate-change-and-energy-transition-law.

Cluster collaboration. Profile.clustercollaboration.eu/profile/cluster-organisation/25610a7e-03ed-400d-912f-d46a43b13b18. (n.d.). Retrieved June 2, 2025, from https://profile.clustercollaboration.eu/profile/cluster-organisation/25610a7e-03ed-400d-912f-d46a43b13b18.

CorPower Ocean—Wave Power. To Power the Planet. (2021, December 13). Retrieved from https://corpowerocean.com/.

CSIS. (2023). *Is the Global Workforce Ready for the Energy Transition?* Link.

Deep Ocean Mission | India Science, Technology & Innovation—ISTI Portal. (n.d.). Retrieved June 2, 2025, from https://www.indiascienceandtechnology.gov.in/st-visions/national-mission/deep-ocean-mission.

EMEC: European Marine Energy Centre. (2024). Retrieved June 2, 2025, from https://www.emec.org.uk/.

Europe hydropower regional profileHydropower in Europe. (2025). Retrieved April 15, 2025, from https://www.hydropower.org/region-profiles/europe.

Financing Climate Futures | OECD. (2025). Retrieved June 2, 2025, from https://www.oecd.org/en/publications/financing-climate-futures_9789264308114-en.html.

Frid, C., Andonegi, E., Depestele, J., Judd, A., Rihan, D., Rogers, S., & Kenchington, E. (2011). The environmental interaction of tidal and wave energy generation devices. *Environmental Impact Assessment Review, 32*, 133–139. https://doi.org/10.1016/j.eiar.2011.06.002

Gemechu, E., & Kumar, A. (2022). A review of how life cycle assessment has been used to assess the environmental impacts of hydropower energy. *Renewable and Sustainable Energy Reviews, 167*, Article 112684. https://doi.org/10.1016/j.rser.2022.112684

Ghandehariun, S., Ghandehariun, A. M., & Ziabari, N. B. (2024). Complementary assessment and design optimization of a hybrid renewable energy system integrated with pumped hydro energy storage with natural intake. *Renewable Energy*. Retrieved from https://api.semanticscholar.org/CorpusID:269350046.

Global Maritime Forum. (2023). *Green Jobs and Maritime Decarbonisation*. Link.

Gonzalez-Salazar, M., & Poganietz, W. R. (2022). Making use of the complementarity of hydropower and variable renewable energy in Latin America: A probabilistic analysis. *Energy Strategy Reviews*. Retrieved from https://api.semanticscholar.org/CorpusID:252804092.

G-res Tool. (2025). Retrieved June 2, 2025, from https://www.grestool.org/.

Grill, G., Lehner, B., Lumsdon, A. E., MacDonald, G. K., Zarfl, C., & Reidy Liermann, C. (2015). An index-based framework for assessing patterns and trends in river fragmentation and flow regulation by global dams at multiple scales. *Environmental Research Letters, 10*(1), Article 015001. https://doi.org/10.1088/1748-9326/10/1/015001

GWEC. (2023). *Global Offshore Wind Report*. Link.

Helseth, A., Melo, A. C. G., Ploussard, Q., Mo, B., Maceira, M. E. P., Botterud, A., & Voisin, N. (2023). Hydropower scheduling toolchains: Comparing experiences in Brazil, Norway, and USA and implications for synergistic research. *Journal of Water Resources Planning and Management*. Retrieved from https://api.semanticscholar.org/CorpusID:258682982.

Home. (n.d.). HIE. Retrieved June 2, 2025, from https://www.waveenergyscotland.co.uk/.

Home—KineticNRG | Green Energy Supplier. (n.d.). Retrieved June 2, 2025, from https://kineticnrg.com.au/.

Home—OceanHub Africa. (2025). Retrieved June 2, 2025, from https://www.oceanhub.africa/.

Horizon 2020-Funded EPIC Project Releases Policy Recommendations for Europe and New Zealand's AI Collaboration Future | News | CORDIS | European

Commission. (2025). Retrieved May 21, 2025, from https://cordis.europa. eu/article/id/125729-horizon-2020funded-epic-project-releases-policy-rec ommendations-for-europe-and-new-zealands-a.

Hydropower Sustainability Assessment Protocol. (2025). Retrieved May 20, 2025, from https://www.hydropower.org/publications/hydropower-sustainability-assessment-protocol.

IEC TS 62600-2:2019. (2025). Retrieved June 2, 2025, from https://webstore. iec.ch/en/publication/62399.

International Energy Agency (IEA). (2023). *Ocean Energy Systems*. Link.

International Renewable Energy Agency (IRENA). (2022). *Innovation Outlook: Ocean Energy*. Link.

IPCC. (2023). *AR6 Synthesis Report: Climate Change 2023*. IPCC. Retrieved from https://www.ipcc.ch/report/ar6/syr/.

IPCC Working Group I Summary for Policymakers in the UN Official Languages—IPCC. (2025). Retrieved March 7, 2024, from https://www. ipcc.ch/2022/11/16/ipcc-wg1-ar6-summary-for-policymakers-un-official-languages/.

IRENA. (2023). *Renewable energy and jobs: Annual review 2023*. Link.

Jiang, J., Ming, B., Huang, Q., Guo, Y., Shang, J., Jurasz, J., & Liu, P. (2023). A holistic techno-economic evaluation framework for sizing renewable power plant in a hydro-based hybrid generation system. *Applied Energy, 348*, Article 121537. https://doi.org/10.1016/j.apenergy.2023.121537

Khedkar, K., Nangia, N., Thirumalaisamy, R., & Bhalla, A. P. S. (2021). The inertial sea wave energy converter (ISWEC) technology: Device-physics, multiphase modeling and simulations. *Ocean Engineering, 229*, Article 108879. https://doi.org/10.1016/j.oceaneng.2021.108879

Łącka, I. (2023). The role of green energy in the economic growth of the world. In I. Bąk & K. Cheba (Eds.), *Green energy: Meta-analysis of the research results* (pp. 41–57). Springer International Publishing. https://doi.org/10. 1007/978-3-031-12531-7_3

Marine energy capacity worldwide 2024. (2024). Statista. Retrieved June 3, 2025, from https://www.statista.com/statistics/476267/global-capacity-of-marine-energy/.

Marine Energy Conferences 2025 | Marine Energy Conference 2025 | Marine Energy Events 2025 | Marine Energy Meetings 2025. (n.d.). Retrieved May 21, 2025, from https://magnusconferences.com/green-chemistry/program/ scientific-sessions/marine-energy.

Mazi, K., Koussis, A. D., Lykoudis, S., Vitantzakis, G., Dimitriadis, P., Kappos, N., Psiloglou, B. E., Katsanos, D., Koletsis, I., Rozos, E., & Kopania, T.

(2021). *HYDRO-NET: Hydro-telemetric Network for surface waters—Innovations and Prospects*. Retrieved from https://api.semanticscholar.org/Cor pusID:236765223.

MeyGen Tidal Energy Project | Tethys. (2025). Retrieved June 2, 2025, from https://tethys.pnnl.gov/project-sites/meygen-tidal-energy-project.

Mikhailov, V. E., Ivanchenko, I. P., & Prokopenko, A. N. (2021). Modern state of hydropower and construction of hydro turbines in Russia and abroad. *Thermal Engineering, 68*(2), 83–93. https://doi.org/10.1134/S00 4060152102004X

Ministry of Industries, Mines and Energy—Gas to Power Project. (2025). Retrieved June 2, 2025, from https://www.mme.gov.na/petroleum/kud ugas/.

Morani, M. C., Carravetta, A., Fecarotta, O., & Montillo, R. (2024). Energy assessment of water networks based on new performance indicators. *Engineering Proceedings, 69*(1), 128. https://doi.org/10.3390/engproc20240 69128

Multon, B. (2012). *Marine renewable energy handbook*. Retrieved from https://www.amazon.com/Marine-Renewable-Energy-Handbook-Bernard/dp/184 8213328.

Normandie Hydroliennes—Projet NH1. (2025). Retrieved June 2, 2025, from https://normandiehydroliennes.fr/en/le-projet-nh1/.

Nugraha, A. T., & Priyambodo, D. (2021). Design of hybrid portable underwater turbine hydro and solar energy power plants: Innovation to use underwater and solar current as alternative electricity in Dusun Dongol Sidoarjo. *Journal of Electronics, Electromedical Engineering, and Medical Informatics*. Retrieved from https://api.semanticscholar.org/CorpusID:237 646092.

Ocean Energy Systems (OES). (2023). *Annual Report*. Link.

Ocean Power Products—SeaRAY & StingRAY AOPS | C-Power. (n.d.). Retrieved June 2, 2025, from https://cpower.co/products/.

Offshore Renewable Energy Development Plan (OREDP). (2025). Gov.Ie. Retrieved June 2, 2025, from https://gov.ie/en/department-of-the-enviro nment-climate-and-communications/publications/offshore-renewable-ene rgy-development-plan-oredp/.

Offshore Wind Energy Impacts on Fishing Communities | NOAA Fisheries. (n.d.). Retrieved May 20, 2025, from https://www.fisheries.noaa.gov/topic/ offshore-wind-energy/evaluating-impacts-to-fisheries.

Open Access Government. (2024). *Sustainable marine technologies and innovation—TFI Marine SeaSpring*. https://doi.org/10.56367/oag-041-11264

Phanindra, K., Bavirisetti, K., Balaji, K., Krishna, P. D. S., & Ganesh, K. M. (2024). Renewable energy technologies: Advances and challenges. *Deleted Journal, 4*(6), 395–397. https://doi.org/10.62225/2583049x.2024.4.6.3451

Renewable Energy Integration Demonstrator—Singapore Flagship Programme. (2025). Energy Research Institute @ NTU. Retrieved June 2, 2025, from https://www.ntu.edu.sg/erian/research-focus/flagship-programmes/renewable-energy-integration-demonstrator-singapore.

RheEnergise High-Density Hydro—How it works. (2025). Retrieved June 2, 2025, from https://www.rheenergise.com/how-it-works.

RivGen® Power System & Integrated Microgrid Solutions. (2025). *ORPC.* Retrieved June 2, 2025, from https://orpc.co/rivgen-power-system-integrated-microgrid-solutions-2/.

Satymov, R., Bogdanov, D., Dadashi, M., Lavidas, G., & Breyer, C. (2024). Techno-economic assessment of global and regional wave energy resource potentials and profiles in hourly resolution. *Applied Energy, 364*, Article 123119. https://doi.org/10.1016/j.apenergy.2024.123119

Seriño, M. N. V. (2021). Energy security through diversification of non-hydro renewable energy sources in developing countries. *Energy & Environment, 33*, 546–561.

Shove, E., & Walker, G. (2010). Governing transitions in the sustainability of everyday life. *Research Policy, 39*(4), 471–476.

South Test Site—PacWave. (2025). Retrieved June 2, 2025, from https://pacwaveenergy.org/south-test-site/.

Special Report on the Ocean and Cryosphere in a Changing Climate. (n.d.). Retrieved June 2, 2025, from https://www.ipcc.ch/srocc/.

Stiglitz, J. E. (2002). *Globalization and its discontents.* W.W. Norton & Company.

Summary Tables. (2024). *Global Energy Monitor.* Retrieved April 15, 2025, from https://globalenergymonitor.org/projects/global-hydropower-tracker/summary-tables/.

Supervisory Control and Data Acquisition System—An overview | ScienceDirect Topics. (2025). Retrieved June 2, 2025, from https://www.sciencedirect.com/topics/computer-science/supervisory-control-and-data-acquisition-system.

Swansea Bay Tidal Lagoon (SBTL) | Tethys. (2025). Retrieved June 2, 2025, from https://tethys.pnnl.gov/project-sites/swansea-bay-tidal-lagoon-sbtl.

Tafur, V. M. (2011). *Water Law, Mining and Hydro-Energy Conflicts in South America: Tales from the Andes and Patagonia.* Retrieved from https://api.semanticscholar.org/CorpusID:140136574.

Technology case study: Sihwa Lake tidal power station. (2025). Retrieved June 2, 2025, from https://www.hydropower.org/blog/technology-case-study-sihwa-lake-tidal-power-station.

The North Seas Energy Cooperation. (2025). Retrieved June 2, 2025, from https://energy.ec.europa.eu/topics/infrastructure/high-level-groups/north-seas-energy-cooperation_en.

Thorson, J., Matthews, C. E., Lawson, M. J., Hartmann, K., Anwar, M. B., & Jadun, P. (2022). *Unlocking the potential of marine energy using hydrogen generation technologies.* Retrieved from https://www.nrel.gov/docs/fy22osti/82538.pdf.

TIGER: Tidal Stream Industry Energiser project. (2025). Retrieved June 2, 2025, from https://interregtiger.com/.

United Nations. (2023). *Renewable Energy—Powering a Safer Future.* Link.

Vaidya, R. A., Molden, D. J., Shrestha, A. B., Wagle, N., & Tortajada, C. (2021). The role of hydropower in South Asia's energy future. *International Journal of Water Resources Development, 37,* 367–391.

Water Framework Directive—European Commission. (2025, May 21). Retrieved from https://environment.ec.europa.eu/topics/water/water-framework-directive_en.

Wheatley, M. (2024). *Advancements in renewable energy technologies: A decade in review.* https://doi.org/10.70389/pjs.100013

Wiloso, E. I., & Heijungs, R. (2013). Key issues in conducting life cycle assessment of bio-based renewable energy sources. In A. Singh, D. Pant, & S. I. Olsen (Eds.), *Life cycle assessment of renewable energy sources* (pp. 13–36). Springer. https://doi.org/10.1007/978-1-4471-5364-1_2

Windfloat Atlantic | Offshore wind energy. (2025). Windfloat Atlantic. Retrieved June 2, 2025, from https://www.windfloat-atlantic.com/.

You need these skills to get that green job. (2022, March 14). World Economic Forum. Retrieved from https://www.weforum.org/agenda/2022/03/green-skills-for-future-jobs/.

5

Circular Water Economy and the Water-Energy Nexus

Robert C. Brears

Abstract This chapter explores the intersection of the circular water economy and the water-energy nexus, presenting a systems-based approach to resource management that emphasizes efficiency, resilience, and sustainability. It introduces the 5Rs framework, Reduce, Reuse, Recycle, Recover, and Restore, as a practical guide for transforming linear water systems into circular models that reduce energy demand, recover valuable resources, and regenerate natural capital. The chapter outlines the environmental, economic, and regulatory drivers behind the shift to circularity, presenting global best practices through utility-led case studies. By integrating decentralized infrastructure, digital technologies, and nature-based solutions, the circular water economy aligns water service delivery with climate mitigation and adaptation objectives, contributing to long-term water and energy security.

Keywords Circular water economy · Water-energy nexus solutions · Sustainable water management · Energy-efficient water systems · Resource recovery in wastewater

© The Author(s), under exclusive license to Springer Nature Switzerland AG 2025
Y. Ermolaeva et al., *Rethinking Water and Energy for a Sustainable Future*, Palgrave Studies in Climate Resilient Societies, https://doi.org/10.1007/978-3-032-04485-3_5

Introduction

Water and energy are fundamentally interconnected. Every stage of the water cycle, from abstraction and treatment to distribution and disposal, requires energy, while many energy systems depend on water for cooling, processing, and generation. As climate change, urbanization, and resource constraints intensify, the traditional linear model of water management, characterized by extract, use, and dispose, is proving increasingly unsustainable. In response, the circular water economy has emerged as a systems-based approach that reimagines water as a renewable resource within a closed-loop system.

The circular water economy seeks to design out externalities, keep resources in continuous use, and regenerate natural systems. It aligns closely with the objectives of climate resilience and energy efficiency by reducing dependency on high-energy infrastructure, lowering greenhouse gas emissions, and preserving natural capital. Central to this approach is the 5Rs framework: Reduce, Reuse, Recycle, Recover, and Restore. These five principles offer a practical roadmap for utilities, cities, and industries to transition toward more sustainable and integrated water and energy systems.

This chapter explores each of the 5Rs in detail, illustrating their application through global best practices. It highlights how the circular water economy can serve as a foundation for long-term water security while reducing energy demand and supporting broader environmental and economic goals.

Definition of the Circular Water Economy

The circular water economy is a systems-based approach to managing water resources that seeks to design out externalities, keep resources in continuous use, and regenerate natural capital. In contrast to the traditional linear model of extract-use-dispose, the circular water economy reimagines water as part of a closed-loop system that maximizes efficiency, minimizes waste, and promotes long-term sustainability in terms of environmental, economic, and social benefits.

Designing out externalities in the circular water economy means minimizing the negative impacts associated with water use and wastewater treatment. This involves optimizing the use of energy, minerals, and chemicals in water systems to reduce inputs and emissions. It also means minimizing consumptive use of water by adopting solutions that either reduce water use or deliver the same outcomes without using water at all. By integrating water services more closely with other urban systems, such as energy, transport, and food, the circular water economy enables more effective resource planning, avoids redundancies, and lowers overall environmental impact.

Keeping resources in use focuses on extending the value and utility of resources embedded in water systems. This includes optimizing water use and reuse, as well as recovering and reusing energy, nutrients, and chemicals wherever possible. Water is no longer viewed as a one-time-use commodity but as a service that can be cycled repeatedly. Additionally, the circular model maximizes value creation across interfaces where water systems intersect with agriculture, energy production, industry, and urban infrastructure. This fosters synergy and innovation, particularly in areas such as decentralized reuse, nutrient recovery, and co-located systems that integrate water and energy functions.

Regenerating natural capital aims to restore and enhance the ecosystems that underpin water availability and quality. This involves ensuring environmental flows by reducing both consumptive and non-consumptive water uses, and minimizing pollution through improved effluent quality and source control. By preserving wetlands, aquifers, and watersheds and reducing human disruption to natural water cycles, the circular water economy supports the long-term resilience of hydrological systems. Natural capital is treated not only as a constraint but as an asset to be protected and enhanced through thoughtful water governance and integrated planning.

Beyond environmental goals, the circular water economy plays a crucial role in mitigating and adapting to climate change. Traditional water supply and wastewater treatment processes are energy-intensive, often relying on fossil fuels for power. In contrast, circular practices enable energy efficiency improvements and the generation of renewable energy through biogas from anaerobic digestion, heat recovery from wastewater, and hydropower. These measures contribute to lower greenhouse gas emissions while improving operational performance. Utilities investing in such technologies often experience short payback periods and reduced operational costs, which can lead to lower tariffs for customers and increased water availability through leakage reduction.

Ultimately, the circular water economy enhances climate resilience. By ensuring that water systems can cope with disruptions, anticipate future variability, and continue to provide essential services, the model supports sustainable development. Resilience encompasses both the physical infrastructure and institutional capacity required to adapt to change, protect natural systems, and ensure long-term access to water for communities and ecosystems alike.[1,2]

Drivers of the Circular Water Economy

The transition toward a circular water economy is being driven by a combination of climatic, economic, regulatory, and infrastructural pressures that challenge the long-term sustainability of conventional, linear water systems. These drivers highlight the limitations of the extract-use-dispose model and encourage the adoption of circular strategies that optimize water use, recover embedded resources, and regenerate natural capital.

Climate change and extreme weather events are among the most significant drivers accelerating the shift to circular water practices. Rising temperatures and more frequent heatwaves increase water demand for cooling, irrigation, and recreational purposes, while also deteriorating water quality through the proliferation of pathogens, algae, and invasive species. Reduced snowmelt inflows, lower rainfall, and more prolonged droughts diminish the quantity and quality of water available, creating tensions between consumptive and environmental uses. Additionally, power outages triggered by extreme weather can disrupt water and wastewater services, exposing the vulnerability of linear infrastructure systems.

More intense rainfall and flash flooding place further stress on urban drainage and sewerage infrastructure. Combined sewer overflows, runoff contamination after dry periods, and direct damage to water assets from storm events highlight the need for more adaptive and integrated water management approaches. Restoring natural hydrological functions and implementing nature-based solutions are increasingly recognized as crucial strategies for managing these risks.

The energy-intensiveness of water systems represents another key driver. Water extraction, treatment, conveyance, and wastewater processing consume substantial amounts of energy, often sourced from fossil fuels, resulting in significant greenhouse gas emissions. As climate change continues to alter the quantity and quality of accessible water sources, utilities are forced to draw water from more distant or degraded sources, increasing energy use and operational costs. The circular water economy presents opportunities to decouple water services from fossil energy dependence through energy efficiency measures, on-site renewable generation, and the recovery of biogas and heat from wastewater streams.

Economic growth and urbanization further intensify water-related pressures. As populations expand and urban areas grow, the demand for water services increases alongside per capita consumption, particularly in higher-income settings. This expansion often outpaces the capacity of existing infrastructure, much of which is aging and increasingly costly to maintain. Circular models that emphasize reuse, recycling, and localized resource recovery can help alleviate the demand for new infrastructure while extending the lifespan and enhancing the performance of existing systems.

Additionally, non-climatic drivers such as rising environmental standards, more stringent water quality regulations, and the integration of ESG (environmental, social, and governance) principles into public and private sector investment frameworks are accelerating the shift to circularity. Water utilities are increasingly expected to demonstrate environmental responsibility, resilience, and transparency in their operations. At the same time, customer expectations are evolving, with growing interest in sustainable service delivery, transparency in resource use, and affordability.

The circular water economy offers a practical response to this convergence of challenges by creating systems that are more efficient, resilient, and regenerative. These drivers underscore the urgency of adopting the 5Rs framework, Reduce, Reuse, Recycle, Recover, and Restore, as a foundation for transforming water services and aligning them with the broader goals of climate resilience, economic sustainability, and environmental protection.[3,4,5]

The 5Rs Framework in the Circular Water Economy

To operationalize the circular water economy in practice, a structured and replicable approach is necessary, one that enables water authorities, utilities, industries, and communities to integrate circularity into water service planning and operations. The 5Rs framework, Reduce, Reuse, Recycle, Recover, and Restore, offers a comprehensive and systemic model for achieving this. Each component targets a distinct intervention point in the water cycle, contributing to more efficient water use, reduced energy demand, and enhanced environmental sustainability.

The 5Rs are not meant to function in isolation. Instead, they are interconnected principles that guide actions across the full lifecycle of water and its embedded resources, including energy, nutrients, and materials. Together, they form a circular pathway that shifts water systems from linear, consumption-driven models toward resource-conscious, regenerative systems. Importantly, this approach also responds directly to the increasing interdependence between water and energy. In both urban and rural contexts, water services, abstraction, treatment, conveyance, distribution, and discharge are energy-intensive. Conversely, many forms of energy production depend on water, for cooling, steam generation, or as an input in renewable energy technologies. As such, each of the 5Rs contributes to reducing pressure on the water-energy nexus by either limiting demand, improving efficiency, or substituting conventional resource use.

Reduce focuses on minimizing water demand and the associated energy use at the source. This includes promoting water-use efficiency through smart metering, leak detection, pricing incentives, and behavioral change. Reducing consumption at the outset means less water needs to be treated and pumped, thereby reducing operational energy requirements and infrastructure strain.

Reuse encourages the use of water more than once before it is discharged from the system. This often occurs at the household, building, or community scale, where rainwater or graywater is reused for non-potable purposes, such as irrigation, toilet flushing, or industrial cooling. Reuse reduces the demand for treated freshwater and limits the volume of wastewater generated, both of which lower energy needs in supply and treatment.

Recycling involves treating wastewater to a quality suitable for further use and redirecting it for agricultural, industrial, or urban applications. Recycling reduces the extraction of freshwater sources and can replace energy-intensive alternatives such as desalination or long-distance transfers. It also supports local water security by diversifying supply options.

Recover refers to the extraction of embedded resources, such as nutrients, biogas, and thermal energy, from wastewater. These recovered materials can be reused in agriculture or fed back into the energy grid, reducing the reliance on external, carbon-intensive inputs. Recovery can transform treatment plants from energy consumers into net producers.

Restore promotes the regeneration of natural water systems to enhance their capacity to regulate, filter, and store water. Healthy ecosystems reduce the need for artificial water treatment and conveyance, lowering energy inputs while preserving biodiversity and hydrological function.

By applying the 5Rs framework in concert, stakeholders can design, implement, and manage water systems that are not only more efficient and less polluting but also more resilient and adaptable in the face of climate change and resource constraints.[6,7]

Reduce: Enhancing Efficiency to Lower Water-Energy Nexus Pressures

Reducing water demand is a crucial first step in operationalizing the circular water economy and alleviating pressure on the interconnected water-energy nexus. By decreasing the volume of water extracted, treated, distributed, and disposed of, energy use across the entire water cycle is simultaneously reduced. This integrated benefit highlights why "reduce" serves as the foundation of circular water management practices. In this context, water authorities, utilities, and end users can implement multiple demand-reduction strategies to improve efficiency and conserve both water and energy resources.

Water conservation efforts begin with promoting best management practices tailored to local water availability, climate conditions, and user behavior. In agriculture, drip irrigation systems deliver water directly to plant roots, minimizing evaporation and runoff. This precision not only reduces water consumption but also cuts the energy required for pumping and transporting large volumes of water across fields. Since irrigation accounts for a significant share of global water withdrawals and energy use in the water sector, efficiency improvements in this area offer substantial co-benefits for both water and energy systems.

In urban and industrial settings, reducing demand is closely linked to the deployment of smart metering and water-efficient fixtures. Smart meters enable real-time monitoring of water use, allowing consumers and utilities to detect leaks, adjust consumption patterns, and implement conservation measures. The granular data provided by these systems also supports demand forecasting and infrastructure planning, further reducing the risk of overcapacity and unnecessary energy expenditure in water treatment and distribution.

Pricing mechanisms can also play a central role in reducing water consumption and easing associated energy demands. Tiered pricing structures, where unit prices increase with higher levels of use, encourage conservation by creating a financial incentive for users to reduce consumption. When paired with public awareness campaigns and technical assistance programs, these pricing strategies can shift user behavior toward more sustainable patterns without compromising access to essential water services.

Importantly, reducing non-revenue water, water that is produced but lost before reaching end users due to leaks, illegal connections, or metering inaccuracies, offers another significant opportunity. By investing in infrastructure upgrades, pressure management, and leakage detection, water utilities can reduce physical water losses. Since every unit of lost water represents wasted energy for treatment and transport, reducing these losses directly contributes to decreasing the energy intensity of water services.

The impact of reduction strategies on the water-energy nexus is further amplified when considering indirect energy savings. For example, reducing household water use lowers the volume of wastewater generated, thereby decreasing the load on sewer networks and treatment plants. This cascade effect helps limit energy consumption in both the supply and sanitation sides of the system. Additionally, reducing demand can delay or eliminate the need for energy-intensive capital investments, such as desalination plants or new long-distance conveyance infrastructure.

Overall, reducing water consumption through targeted efficiency measures is essential for creating a circular water economy that simultaneously reduces the energy demand. By prioritizing reduction at the source, water authorities can deliver a range of benefits that enhance resource security, operational efficiency, and environmental sustainability.[8,9,10,11,12,13]

Case Study: Dubai Electricity and Water Authority—Smart Water Metering System

The Dubai Electricity and Water Authority (DEWA) has successfully achieved 100% installation of smart water meters across Dubai, with over one million units deployed by mid-2024. These meters are part of DEWA's broader smart grid and digital transformation initiatives aimed at enhancing resource management, operational efficiency, and customer engagement.

Smart water meters provide customers with real-time and historical data on their water usage, enabling them to make informed decisions about reducing consumption and detecting leaks. Through DEWA's Smart Living platform, users can compare their consumption with that of similar households, receive high-usage alerts, activate away mode, and self-assess their usage via an interactive tool. These features empower customers to proactively manage their water consumption without requiring direct contact with DEWA.

DEWA manages smart meter data through a secure, automated infrastructure that integrates with its Advanced Metering Infrastructure (AMI) and SAP billing system. Meters are monitored remotely every 15 minutes through the Smart Meters Analysis and Diagnosis Centre. Additionally, DEWA has developed Hydro Insight, an internally built monitoring system that can detect anomalies across the network within an hour.

The smart metering initiative has made a significant contribution to resource efficiency. Between 2018 and 2023, DEWA reduced water losses from 7.06% to 4.6%, saving an estimated 12.5 billion gallons of water and AED 451.3 million. This positions Dubai among the global leaders in water network performance.

Smart meters also enable accurate billing and reduce unaccounted-for water. With automated meter reading, customers receive consistent updates and billing based on actual consumption, enhancing transparency and satisfaction.

The deployment supports Dubai's Economic Agenda D33 by leveraging Fourth Industrial Revolution technologies to drive sustainability and innovation. DEWA's smart water infrastructure reflects the emirate's commitment to digital transformation, net-zero ambitions, and improved quality of life for residents.[14]

Reuse: Maximizing Local Water Availability to Reduce Energy Demand

Reuse is the second pillar of the circular water economy, playing a critical role in reducing pressure on the water-energy nexus. By using water more than once before it exits the system, communities and utilities can reduce the need to extract, treat, and transport additional volumes of freshwater, each of which is energy-intensive. Reuse strategies, especially those that require little or no treatment, offer a cost-effective and low-energy solution to increasing water stress and growing urban demands.

At the household and building scales, reuse can be achieved by installing systems that harvest greywater and rainwater for non-potable applications. Gray water, wastewater from showers, sinks, and washing machines, can be collected and reused for purposes such as toilet flushing, landscape irrigation, or cleaning. Rainwater harvesting systems capture runoff from rooftops and other surfaces, storing it for later use, typically for outdoor watering or industrial cooling. These reuse practices reduce reliance on centralized water supply systems and the energy required to extract, treat, and deliver potable water over long distances.

In commercial and institutional settings, reuse systems are increasingly being integrated into building design and site operations. Facilities such as hotels, schools, and manufacturing plants can implement decentralized reuse systems to meet their non-potable water demands internally. This not only lowers their overall water consumption but also eases peak demand on municipal systems. From an energy perspective, these practices avoid the upstream energy costs associated with supplying high-quality water for applications that do not require it.

The benefits of reuse also extend to municipal water utilities. By encouraging decentralized reuse systems and supporting policies that allow and regulate non-potable reuse, utilities can delay or avoid expanding supply infrastructure, which often involves significant capital investment and long-term energy commitments. Moreover, localized reuse can reduce the need to lift and pump water to higher elevations, a process that consumes considerable energy in many urban water systems.

Reuse also helps to manage wastewater volumes by reducing the amount of water that needs to be treated and discharged. Smaller wastewater flows reduce the energy required for pumping and treating effluent, particularly during peak wet weather events when energy consumption can increase significantly. This is particularly valuable in cities with combined sewer systems, where stormwater and wastewater are collected together and place a heavy burden on treatment plants during heavy rains.

Importantly, reuse strategies must be implemented with proper safeguards and governance structures to ensure public health and environmental protection. Simple filtration and disinfection steps can be added to reuse systems as needed, depending on the intended use. With clear guidelines, training, and oversight, decentralized reuse systems can scale safely and effectively across residential, commercial, and industrial sectors.

By integrating reuse practices into water service planning, authorities and end users can significantly reduce their dependence on energy-intensive water supply infrastructure. This alignment with circular economy principles helps mitigate the dual pressures of water scarcity and rising energy demand, contributing to more resilient and efficient urban systems.[15,16,17,18,19,20]

Case Study: San Francisco Public Utilities Commission—Onsite Water Reuse Grant Program

The San Francisco Public Utilities Commission (SFPUC) administers the Onsite Water Reuse Grant Program to promote sustainable water practices and reduce reliance on potable water across the city. This initiative encourages the installation of systems that collect, treat, and reuse alternative water sources for non-potable uses, including toilet flushing, irrigation, and cooling tower makeup.

The grant program supports three categories of projects: voluntary installations, mandated installations that exceed baseline compliance under the city's Nonpotable Water Ordinance, and the onsite treatment and reuse of brewery process water. Eligible alternate sources include rainwater, stormwater, graywater, foundation drainage, air conditioning condensate, and blackwater.

To qualify for funding, projects must demonstrate a significant reduction in SFPUC water usage over a ten-year period. Grant tiers are defined by water offset thresholds: $200,000 for projects saving at least 3.6 million gallons, $500,000 for those saving 8 million gallons, and $1 million for projects achieving a 24 million gallon offset. Projects categorized as "Above and Beyond" must surpass mandatory baseline compliance to be eligible.

Applicants must submit proposals detailing how their project will achieve the required offsets through activities such as installing water collection systems, treatment infrastructure, and storage facilities for treated water. Grant funding is awarded on a first-come, first-served basis, contingent upon eligibility and availability of funds. Applications are reviewed individually and must comply with program rules and water offset calculations specific to each project category.

Grant funding is restricted to non-mandated installations unless the project meets the Above and Beyond criteria. Speculative water-saving projects or those undertaken solely to meet ordinance requirements are excluded. The grant program supports SFPUC's long-term water resilience objectives. It aligns with broader goals established under the Water System Improvement Program, enabling the city to diversify and secure its water supply through proactive onsite reuse.[21]

Recycle: Treating and Reintegrating Water to Reduce Systemic Energy Demand

Recycling, in the context of the circular water economy, refers to the process of treating wastewater to a level that makes it suitable for further use and then reintegrating it into the system. Unlike reuse, which may involve minimal or no treatment, recycling requires treatment to meet quality standards appropriate for specific applications. This added step, however, brings significant long-term benefits by decreasing freshwater extraction, minimizing environmental discharge, and reducing energy consumption across both water and energy systems.

Municipal wastewater treatment plants play a central role in advancing recycling practices. When treated effluent is redirected for non-potable applications, such as irrigating parks, golf courses, and farmland, or cooling in industrial processes, demand for potable water is reduced. This means less water needs to be abstracted, treated, and pumped to users, all of which are energy-intensive processes. Furthermore, supplying recycled water locally avoids the energy costs associated with transporting water from distant sources.

The potential for energy savings is particularly significant in areas where water must be transported over long distances or lifted to higher elevations. In such contexts, substituting recycled water for freshwater can markedly lower overall system energy use. This makes recycling not only a water conservation strategy but also a key efficiency measure in energy management.

Housing developments and industrial parks are increasingly incorporating on-site recycling systems. These systems treat wastewater generated on-site to standards suitable for secondary use. Treated water can then be reused for landscaping, toilet flushing, or process water. By treating and cycling water within the boundaries of a development or facility, these systems create localized loops that reduce reliance on central infrastructure. This decentralization offers operational flexibility and energy efficiency, particularly in rapidly urbanizing or water-scarce regions.

While the treatment process itself requires energy, the overall balance can still favor recycling from an energy perspective. Technological advances in membrane filtration, UV disinfection, and energy-efficient aeration systems have made recycling increasingly viable. In many cases, the energy used for treating water to a non-potable standard is substantially less than the energy required to extract, treat, and deliver new supplies from natural sources. Moreover, recycling can be integrated with energy recovery systems, such as capturing heat from treated effluent, thereby further enhancing the facility's energy profile.

Recycling also reduces the environmental and energy costs associated with discharging treated wastewater into rivers, lakes, or oceans. When this discharge is minimized, less energy is needed for environmental monitoring and pollution control. Additionally, the risk of downstream users requiring further treatment is reduced, resulting in indirect energy savings throughout the wider catchment.

Implementing water recycling on a broad scale requires enabling regulations, public acceptance, and investment in infrastructure. However, where conditions permit, it presents a practical, circular solution for managing water and energy demand simultaneously. As urban centers and industrial hubs seek sustainable ways to support growth, recycling offers a viable path to reduce pressure on the water-energy nexus while enhancing supply reliability and resilience.[22,23,24,25,26,27]

Case Study: Northern Adelaide Irrigation Scheme—Expanding South Australia's Recycled Water Future

The Northern Adelaide Irrigation Scheme (NAIS), delivered by SA Water in partnership with the South Australian Government and the Australian Government, represents a significant step forward in climate-independent water resource management for the agricultural sector. NAIS provides secure access to high-quality recycled water, supporting intensive horticulture, floriculture, and advanced food production across the Northern Adelaide Plains.

Through an initial $155.6 million investment, including $110 million from the state and $45.6 million from the federal government, the scheme has enabled the construction of a new wastewater treatment facility producing an additional six gigalitres of fit-for-purpose recycled water annually. The infrastructure includes seasonal balancing storages, a transmission main connecting Bolivar to the region, and an extensive farm-gate distribution network. Customers are supplied through spur lines and connection points designed to meet their specific operational needs.

The NAIS supplies water under long-term, tradeable contracts, up to 45 years in duration, with tightly managed water quality, climate resilience, and stable pricing. This arrangement supports investor confidence, enabling efficient and predictable planning across multiple sectors, including fruit and nut orchards, table and wine grapes, broad-acre cropping, and intensive animal farming.

Currently, NAIS supports up to 3,000 hectares of high-technology agriculture, with the potential to double the current 12 gigalitres of supply as more producers join. This expansion could generate up to 6,000 jobs and attract $2 billion in private investment, injecting an estimated $1 billion annually into the state economy. Contractual terms require customers to be registered legal entities with ABNs and advanced project proposals within the scheme footprint.

As Australia's second-largest recycler of water, SA Water is scaling circular water use through initiatives like NAIS, which not only promote sustainable food production but also demonstrate how public infrastructure investment can catalyze agribusiness growth under changing climate conditions.[28]

Recover: Extracting Value from Water Systems to Reduce Resource and Energy Pressures

Recovery in the circular water economy focuses on extracting useful resources, such as nutrients, thermal energy, and biogas, from water and wastewater streams. Rather than viewing wastewater as a burden to be treated and disposed of, recovery-oriented approaches treat it as a valuable source of secondary resources. By integrating recovery into water services, utilities can reduce reliance on external energy inputs and industrial fertilizers, thereby easing pressure on both water and energy systems.

One of the most established recovery practices is the extraction of nutrients from wastewater. Treatment plants can recover phosphorus and nitrogen, key components of fertilizer, during the treatment process. These nutrients are typically removed to meet effluent quality standards, and with the right technologies, they can be converted into usable products such as struvite or ammonium sulfate. By reintroducing these materials into agricultural supply chains, utilities reduce demand for synthetic fertilizers, which are highly energy-intensive to produce. This not only supports nutrient cycling but also offers cost and environmental savings in the broader energy system.

Energy recovery from wastewater is another growing area of opportunity. Through anaerobic digestion, organic matter in sewage sludge can be broken down to produce biogas, a renewable energy source composed mainly of methane. This biogas can be used to generate heat or electricity, often within the same facility. Many modern wastewater treatment plants now operate as energy-neutral or even energy-positive systems by maximizing biogas recovery. This helps offset the plant's energy demand and reduces dependence on fossil fuel-based electricity.

Thermal energy can also be recovered from wastewater. Even after water has been used in households or industrial facilities, it retains a significant amount of heat. Heat exchangers can capture this energy, particularly in colder climates where temperature differentials are significant, and repurpose it for heating spaces or water. This approach reduces energy consumption in buildings while utilizing a resource that would otherwise be wasted, embedded within daily water flows.

Beyond the plant level, recovery-oriented strategies can be integrated into decentralized systems, such as those found in high-density residential or commercial buildings. These buildings can incorporate systems that recover heat from greywater or small-scale anaerobic digesters that process food and organic waste. When integrated with building energy systems, these technologies reduce the demand for external energy and offer long-term cost savings.

Recovery also contributes to broader sustainability by reducing the volume of residual waste that requires disposal. For example, using sludge as a feedstock for energy production or as a fertilizer reduces the need for landfill space and limits methane emissions from uncontrolled decomposition. Additionally, by designing treatment systems with recovery in mind, utilities can create new revenue streams while more efficiently meeting environmental regulations.

As urban areas grow and the demand for water and energy intensifies, recovery provides a practical and scalable strategy to close resource loops. Incorporating recovery into water services transforms linear systems into integrated networks where waste becomes input, and the demand on external resources, particularly energy, is substantially reduced. This alignment strengthens the resilience of both water and energy infrastructure within a circular economy framework.[29,30,31,32,33,34]

Case Study: HAMBURG WASSER—Producing Biogas from Wastewater

HAMBURG WASSER operates an innovative wastewater-to-energy system that contributes to Hamburg's climate neutrality goals by converting sewage sludge into renewable biogas. At the Köhlbrandhöft treatment plant, biological processes break down organic material in the sludge, producing raw biogas that is primarily composed of methane and carbon dioxide.

This biogas is processed in two on-site gas upgrading facilities (GALA), where it is refined into biomethane. During treatment, non-methane components are removed, and the gas is compressed, liquefied, and odorized before being fed into the city's natural gas network. The biomethane is then used as a climate-friendly alternative to fossil fuels for heating and cooking in homes.

The facilities generate enough biogas each hour to supply up to 1,350 cubic meters of gas to the network, equivalent to the annual energy needs of approximately 5,700 households. This process offsets approximately 12,000 tons of CO_2 emissions per year. To achieve the same CO_2 reduction through forestry would require roughly 1,000 hectares of beech forest, an area comparable in size to Hamburg's Bramfeld district.

A key feature of the system is the integration of climate-conscious infrastructure. One of the plant's sludge storage tanks is covered by a "climate hood," which captures escaping gases and feeds them into the biomethane production line. This pilot project received support from Germany's Federal Ministry for the Environment.

In addition to producing biomethane, the treatment plant meets a significant portion of its electricity demand by utilizing energy recovered from the wastewater process. Surplus heat is supplied to nearby facilities such as the Tollerort container terminal. Once considered energy-intensive infrastructure, the plant now serves as a model of energy efficiency and circular resource management, reflecting HAMBURG WASSER's commitment to sustainable urban utilities.[35]

Restore: Rehabilitating Natural Systems to Alleviate Water-Energy Nexus Pressures

Restoration within the circular water economy refers to the rehabilitation and protection of natural hydrological systems to support long-term water security and the functionality of ecosystems. Unlike engineered infrastructure that often requires constant energy input, healthy natural systems provide water regulation and filtration services passively. By restoring ecosystems such as wetlands, rivers, floodplains, and aquifers, water authorities can reduce reliance on energy-intensive infrastructure, thus decreasing overall stress on the water-energy nexus.

One of the primary restoration strategies involves recharging groundwater systems through managed aquifer recharge (MAR). In areas where groundwater is being depleted faster than it is replenished, MAR projects introduce treated wastewater or captured stormwater into aquifers through infiltration basins, injection wells, or spreading channels. This process helps restore groundwater levels and maintain base flows in rivers, which are crucial for both human consumption and ecological health. Notably, increasing groundwater storage can reduce the need to extract water from distant sources or rely on desalination, both of which are highly energy-intensive processes.

Restoring wetlands and riparian zones also provides essential water regulation functions. Wetlands act as natural sponges, absorbing excess water during floods and slowly releasing it during dry periods. They also filter pollutants, improving downstream water quality and reducing the energy burden on treatment facilities. Riparian vegetation stabilizes riverbanks, reduces erosion, and enhances groundwater recharge by promoting infiltration of water into the soil. These functions contribute to system-wide resilience and reduce the need for artificial flood control, sediment removal, and water purification, all of which consume significant amounts of energy.

Nature-based solutions can also help restore surface water flows by reconnecting rivers with their floodplains or removing barriers that obstruct ecological continuity. These actions improve the natural retention and movement of water in the landscape. By allowing water to slow, spread, and soak into the ground, restoration projects help maintain a more stable hydrological regime. This diminishes the need for mechanical pumping, stormwater diversion, and peak-flow management systems, all of which require energy input.

Urban restoration plays a growing role in circular water strategies. Cities can integrate green infrastructure, such as permeable pavements, rain gardens, green roofs, and urban forests, to restore some of the natural water functions lost due to development. These systems reduce stormwater runoff volumes, enhance local cooling, and recharge shallow aquifers. By limiting runoff, they also reduce the energy required to pump and treat stormwater, as well as decrease the frequency and intensity of combined sewer overflows.

Ecological restoration contributes to energy efficiency not by generating or reclaiming energy, but by reducing demand for it through natural, passive processes. In this way, it restores ecosystem functions as a foundational support to circular water management. It shifts the burden away from high-energy solutions toward self-regulating systems that provide multiple co-benefits, including biodiversity conservation, climate adaptation, and long-term resource sustainability.

By integrating restoration into water service planning, authorities can unlock natural capacities for water storage, purification, and regulation. This reduces the need for centralized, energy-dependent infrastructure and supports a systemic transition toward circular, resilient, and low-impact water and energy management.[36,37,38]

Case Study: City of Melbourne—Fitzroy Gardens Stormwater Harvesting System

The City of Melbourne implemented the Fitzroy Gardens Stormwater Harvesting System to secure a sustainable irrigation source for its 26-hectare heritage-listed parkland. Designed in response to water shortages, droughts, and climate change, the system captures stormwater runoff from a 67-hectare urban catchment. It reuses it to irrigate the gardens, reducing the park's reliance on drinking water by approximately 59%.

Stormwater is diverted from Wellington Parade into the system, where it first enters a gross pollutant trap, removing coarse debris such as leaves and litter. The water then flows into a nine-meter-deep sedimentation chamber, where it settles out fine sands, oils, and hydrocarbons. From there, partially treated water is stored in a four-million-liter underground primary tank.

The treatment process continues above ground in a 241 m^2 biofiltration bed planted with native wetland grasses. This biofilter removes nutrients, including nitrogen and phosphorus. Water then drains under gravity to a secondary one-million-liter storage tank. Before distribution, the water is disinfected using an ultraviolet (UV) treatment system to eliminate bacteria. It is then pumped through an automated irrigation network supplying the gardens and public amenities.

Challenges during implementation included managing contaminated soil, preserving heritage-listed structures, and integrating the system into a public open space. Soil contamination required in-situ management, and the tank placement was carefully designed to avoid tree root systems and heritage buildings.

Initial issues, such as algae crust formation in the biofilter and elevated salinity levels, were resolved through media replacement and flushing. Ongoing maintenance includes sediment removal, filter media replacement, and monitoring of water quality for turbidity, pH, and salinity. The system contributes to the city's Total Watermark Strategy by reducing pollutant loads entering local waterways and supporting water reuse targets, with ongoing maintenance costs estimated at $28,000 annually.[39]

Best Practices

Each circular water strategy, Reduce, Reuse, Recycle, Recover, and Restore, draws on best practices implemented by the utilities featured in the case studies. These practices demonstrate how targeted interventions across water systems can reduce energy use, alleviate pressure on the water-energy nexus, and support long-term sustainability and resilience.

Reduce

Reducing water demand through smart technologies and efficient management practices provides both immediate and lasting relief to the pressures on the water-energy nexus, forming the foundation of a circular water economy. The following best practices, identified from the DEWA's smart metering initiative, offer transferable insights for other locations aiming to reduce water consumption and associated energy demands across integrated systems:

- **Full-Scale Deployment of Smart Water Meters**

  - *Description*: Achieve universal installation of smart meters to monitor water use across all customer categories.
  - *Water-Energy Nexus Link*: Enables immediate detection of leaks and excessive use, reducing the volume of water that must be extracted, treated, and pumped, thus lowering energy consumption.

- **Consumer Empowerment via Digital Engagement Tools**

  - *Description*: Provide users with access to real-time usage data, comparisons with peers, alerts, and consumption self-assessment tools through platforms like Smart Living.
  - *Water-Energy Nexus Link*: Informed consumers can proactively reduce water use, which also lowers energy demand across water supply and wastewater systems.

- **Integration with AMI**

  - *Description*: Link smart meters with centralized digital platforms like SAP and remote diagnostics centers for automated, real-time network management.
  - *Water-Energy Nexus Link*: Enhances operational efficiency and enables targeted maintenance and pressure optimization, thereby minimizing unnecessary energy consumption.

- **Development of In-House Monitoring Tools**

  - *Description*: Develop proprietary platforms, such as Hydro Insight, for rapid anomaly detection within the water network.
  - *Water-Energy Nexus Link*: Quick resolution of leaks and inefficiencies reduces water and energy loss while improving system resilience.

- **Reduction of Non-Revenue Water through Data-Driven Control**

  - *Description*: Use smart metering data to identify and reduce water losses from leaks or inaccurate billing.

– *Water-Energy Nexus Link*: Prevents the waste of treated and transported water, thus conserving embedded energy throughout the system.

Reuse

Reusing alternative water sources for non-potable purposes reduces dependence on energy-intensive potable supply systems and wastewater treatment processes. The following best practices, identified from SFPUC's Onsite Water Reuse Grant Program, offer transferable insights for other locations aiming to expand water reuse while easing pressures on the water-energy nexus:

- **Incentivize Onsite Reuse Through Tiered Grant Structures**

  – *Description*: Offer financial incentives based on verified water savings over time, with escalating grant amounts tied to greater potable water offsets.
  – *Water-Energy Nexus Link*: Drives the implementation of systems that reduce the need for centralized water extraction, treatment, and distribution, thereby reducing energy use at the system level.

- **Support a Broad Spectrum of Alternative Water Sources**

  – *Description*: Encourage the capture and reuse of multiple non-potable sources, including rainwater, stormwater, graywater, foundation drainage, air conditioning condensate, and blackwater.
  – *Water-Energy Nexus Link*: Utilizing diverse on-site water sources reduces the energy footprint associated with long-distance water conveyance and centralized treatment.

- **Encourage Above-and-Beyond Compliance in Regulated Settings**

  – *Description*: Design grant eligibility to prioritize projects that exceed minimum legal requirements under local ordinances, encouraging deeper reductions.

- *Water-Energy Nexus Link*: Promotes more ambitious water-saving systems that deliver compounded energy savings over standard compliance-level installations.

• **Implement First-Come, First-Served Incentive Allocation**

- *Description*: Streamline funding access through transparent, rolling application review processes to encourage early adoption and reduce administrative burdens.
- *Water-Energy Nexus Link*: Accelerates market uptake of reuse systems that help reduce reliance on energy-intensive potable water supplies.

• **Align Grants with Long-Term Resilience and Infrastructure Planning**

- *Description*: Ensure program goals directly support broader water system strategies, such as resilience objectives under regional improvement plans.
- *Water-Energy Nexus Link*: Integrates localized reuse into system-wide planning, reducing the future need for high-energy supply augmentation (e.g., desalination, inter-basin transfers).

Recycle

Recycling water for productive reuse in agriculture and industry reduces pressure on freshwater sources and mitigates the energy demands of sourcing, treating, and transporting potable supplies. The following best practices, identified from South Australia's NAIS, offer transferable insights for other regions seeking to integrate large-scale recycled water into sustainable resource planning:

• **Develop Climate-Independent Recycled Water Supplies**

- *Description*: Establish advanced treatment facilities and infrastructure that convert wastewater into high-quality recycled water suitable for agricultural and industrial use.

- *Water-Energy Nexus Link*: Reduces dependency on freshwater abstraction and energy-intensive water transfers, while enabling the reuse of existing water resources within local systems.

• **Build Scalable, Demand-Responsive Distribution Networks**

- *Description*: Design modular infrastructure, including seasonal storages, transmission mains, and farm-gate spur lines, that can scale to meet evolving customer needs.
- *Water-Energy Nexus Link*: Optimizes delivery efficiency and avoids overbuilt infrastructure, reducing the embedded energy in supply and storage systems.

• **Implement Long-Term, Tradeable Water Contracts:**

- *Description*: Provide users with stable, multi-decade recycled water contracts to ensure supply certainty and encourage private investment.
- *Water-Energy Nexus Link*: Enables users to shift away from groundwater pumping or energy-intensive irrigation alternatives, aligning long-term planning with reduced energy intensity.

• **Target High-Value, High-Tech Agricultural Applications**

- *Description*: Prioritize recycled water use in advanced horticulture and food production systems that can maximize productivity per unit of water.
- *Water-Energy Nexus Link*: Delivers higher returns on energy invested in water recycling by focusing on sectors with efficient water-energy-productivity ratios.

• **Integrate Recycled Water into Regional Economic Development**

- *Description*: Align water recycling initiatives with broader economic and employment goals to attract investment and foster inclusive growth.
- *Water-Energy Nexus Link*: Embeds circular water use in long-term planning, reducing pressure on energy-intensive future supply augmentation while boosting resilience and local sustainability.

Recover

Recovering energy embedded in wastewater streams transforms treatment infrastructure into net-positive contributors to urban energy systems. The following best practices, identified from HAMBURG WASSER's biomethane initiative, offer transferable insights for utilities aiming to harness energy from sludge while reducing the climate and energy intensity of wastewater management:

- **Convert Sewage Sludge into Renewable Biogas**

  - *Description*: Utilize anaerobic digestion at wastewater treatment plants to break down organic material in sludge and generate raw biogas.
  - *Water-Energy Nexus Link*: Captures energy that would otherwise be lost in waste streams, reducing external energy inputs and fossil fuel dependency.

- **Refine and Inject Biomethane into Urban Gas Networks**

  - *Description*: Upgrade raw biogas on-site to grid-quality biomethane, then distribute it via existing gas infrastructure for household or commercial use.
  - *Water-Energy Nexus Link*: Offsets energy demand in heating and cooking, reducing reliance on fossil energy while valorizing water sector waste.

- **Install Emissions-Capturing Infrastructure**

  - *Description*: Implement covers like "climate hoods" on sludge tanks to capture fugitive emissions and redirect them into energy recovery systems.
  - *Water-Energy Nexus Link*: Prevents methane release, a potent greenhouse gas, and reclaims energy, supporting emissions reduction and resource efficiency.

- **Supply Recovered Heat to Adjacent Facilities**

- *Description*: Repurpose surplus heat from the digestion or combustion process to serve nearby infrastructure, such as industrial terminals or public buildings.
- *Water-Energy Nexus Link*: Displaces external heat sources and closes energy loops within the urban system, maximizing thermal energy recovery from wastewater.

● **Power Treatment Operations with Onsite Energy**

- *Description*: Use electricity and heat generated through the biogas process to meet a substantial share of a plant's operational demands.
- *Water-Energy Nexus Link*: Reduces dependence on grid energy and enhances overall system efficiency by transforming wastewater treatment into a self-sustaining process.

Restore

Restoring urban hydrological functions through decentralized stormwater systems enhances environmental quality and reduces demand on energy-intensive potable water supplies. The following best practices, identified from the City of Melbourne's Fitzroy Gardens Stormwater Harvesting System, offer transferable insights for cities seeking to restore ecological flows and reduce pressures on the water-energy nexus:

● **Capture and Treat Urban Stormwater at Source**

- *Description*: Divert runoff from surrounding streets into engineered treatment systems to enable local reuse.
- *Water-Energy Nexus Link*: Reduces reliance on centralized water treatment and long-distance conveyance, cutting associated energy use.

● **Implement Multi-Stage Natural and Engineered Filtration**

- *Description*: Combine gross pollutant traps, sedimentation chambers, and biofiltration beds to remove solids, oils, and nutrients from stormwater sequentially.

- *Water-Energy Nexus Link*: Improves water quality onsite, minimizing the need for energy-intensive chemical or mechanical treatment downstream.

• **Integrate Storage Tanks for Irrigation Supply**

  - *Description*: Use primary and secondary storage tanks to hold treated stormwater for later use in landscape irrigation and public amenities.
  - *Water-Energy Nexus Link*: Substitutes potable water with treated stormwater, reducing the energy footprint of urban green space management.

• **Apply UV Disinfection Before Use**

  - *Description*: Use ultraviolet systems to eliminate pathogens from stored stormwater before distribution.
  - *Water-Energy Nexus Link*: Offers a low-energy alternative to conventional chlorination or advanced disinfection technologies.

• **Design for Long-Term Maintenance and Landscape Compatibility**

  - *Description*: Address soil contamination, heritage protection, and ecological aesthetics through careful integration into public spaces.
  - *Water-Energy Nexus Link*: Ensures sustained operation of nature-based infrastructure that reduces potable water dependency and enhances passive water treatment.

Conclusion

The circular water economy presents a strategic pathway for addressing the growing pressures on water and energy systems. Shifting from linear models to circular approaches enables the integration of efficiency, sustainability, and resilience into water service planning and delivery. The 5Rs framework, Reduce, Reuse, Recycle, Recover, and Restore, provides a practical structure for implementing circularity across different contexts, guiding stakeholders toward minimizing resource use, maximizing value, and enhancing the natural systems that support water availability and quality.

Each element of the 5Rs contributes to reducing the energy intensity of water systems, either by lowering demand, recovering embedded resources, or leveraging nature-based processes. When applied together, these strategies improve operational performance, reduce greenhouse gas emissions, and build adaptive capacity in the face of climate variability. The case studies and best practices highlighted in this chapter demonstrate that circular water solutions are not only technically feasible but also economically and environmentally beneficial.

As water and energy demands continue to grow, adopting circular approaches becomes essential for long-term sustainability. The circular water economy aligns water management with climate action goals, providing a foundation for resource-secure, low-carbon, and resilient urban and regional systems.

Notes

1. R.C. Brears, *Developing the Circular Water Economy* (Cham, Switzerland: Palgrave Macmillan, 2020a).
2. Robert C. Brears, "Circular Water Economy," in *The Palgrave Encyclopedia of Urban and Regional Futures*, ed. Robert C. Brears (Cham: Springer International Publishing, 2022).
3. R.C. Brears, *Urban Water Security* (Chichester, UK; Hoboken, NJ: John Wiley & Sons, 2016).
4. *The Green Economy and the Water-Energy-Food Nexus*, Second ed. (Cham, Switzerland: Springer International Publishing, 2023b).
5. Robert C. Brears, Financing Water Security and Green Growth, (Oxford University Press, 2023c), https://doi.org/10.1093/oso/9780192847843.001.0001. https://doi.org/10.1093/oso/9780192847843.001.0001.
6. Brears, *Developing the Circular Water Economy*.
7. Brears, "Circular Water Economy.".
8. Brears, *Developing the Circular Water Economy*.
9. Brears, "Circular Water Economy.".

10. *Water Resources Management: Innovative and Green Solutions*, 2 ed. (Berlin, Boston: De Gruyter, 2024).
11. Brears, *Urban Water Security*.
12. *The Green Economy and the Water-Energy-Food Nexus*.
13. *Regional Water Security* (Wiley, 2021).
14. DEWA, "Over a Million Smart Water Meters in Dubai with 100% Installation Rate, Enhancing the Water Network's Advanced and Resilient Infrastructure," https://www.dewa.gov.ae/en/about-us/media-publications/latest-news/2024/07/over-a-million-smart-water-meters-in-dubai-with-100.
15. Brears, *Developing the Circular Water Economy*.
16. Brears, "Circular Water Economy.".
17. *Water Resources Management: Innovative and Green Solutions*.
18. Brears, *Urban Water Security*.
19. *The Green Economy and the Water-Energy-Food Nexus*.
20. *Regional Water Security*.
21. SFPUC, "Onsite Water Reuse Grant," https://www.sfpuc.gov/programs/grants/onsite-water-reuse-grant#:~:text=Our%20Onsite%20Water%20Reuse%20Grant,for%20non%2Dpotable%20uses%20such.
22. Brears, *Developing the Circular Water Economy*.
23. Brears, "Circular Water Economy.".
24. *Water Resources Management: Innovative and Green Solutions*.
25. Brears, *Urban Water Security*.
26. *The Green Economy and the Water-Energy-Food Nexus*.
27. *Regional Water Security*.
28. SA Water, "Overview: Nais—the Future of Food Production in Sa," https://www.sawater.com.au/nais/overview.
29. Brears, *Developing the Circular Water Economy*.
30. Brears, "Circular Water Economy.".
31. *Water Resources Management: Innovative and Green Solutions*.
32. Brears, *Urban Water Security*.
33. *The Green Economy and the Water-Energy-Food Nexus*.
34. *Regional Water Security*.
35. Hamburg Wasser, "Biogas Aus Abwasser," https://www.hamburgwasser.de/umwelt/energiegewinnung/biogas.

36. R.C. Brears, *Blue and Green Cities: The Role of Blue-Green Infrastructure in Managing Urban Water Resources* (Springer International Publishing, 2023a).
37. Brears, *Water Resources Management: Innovative and Green Solutions*.
38. R.C. Brears, *Nature-Based Solutions to 21st Century Challenges* (Oxfordshire, UK: Routledge, 2020b).
39. City of Melbourne, "Fitzroy Gardens Stormwater Harvesting System," https://www.melbourne.vic.gov.au/fitzroy-gardens-stormwater-harvesting-system.

References

Brears, R. C. (2016). *Urban water security*. Wiley.
Brears, R. C. (2020a). *Developing the circular water economy*. Palgrave Macmillan.
Brears, R. C. (2020b). *Nature-based solutions to 21st century challenges*. Routledge.
Brears, R. C. (2021). *Regional water security*. Wiley.
Brears, R. C. (2022). Circular water economy. In R. C. Brears (Ed.), *The Palgrave encyclopedia of urban and regional futures* (pp. 193–199). Springer International Publishing.
Brears, R. C. (2023a). *Blue and green cities: The role of blue-green infrastructure in managing urban water resources*. Springer International Publishing.
Brears, R. C. (2023b). *The green economy and the water-energy-food nexus* (2nd ed.). Springer International Publishing.
Brears, R. C. (2023c). *Financing water security and green growth*. Oxford University Press. https://doi.org/10.1093/oso/9780192847843.001.0001
Brears, R. C. (2024). *Water resources management: Innovative and green solutions* (2nd ed.). De Gruyter. https://doi.org/10.1515/9783111028101
City of Melbourne. *Fitzroy Gardens Stormwater Harvesting System*. Retrieved from https://www.melbourne.vic.gov.au/fitzroy-gardens-stormwater-harvesting-system.
DEWA. *Over a Million Smart Water Meters in Dubai with 100% Installation Rate, Enhancing the Water Network's Advanced and Resilient Infrastructure*.

Retrieved from https://www.dewa.gov.ae/en/about-us/media-publications/latest-news/2024/07/over-a-million-smart-water-meters-in-dubai-with-100.

Hamburg Wasser. *Biogas Aus Abwasser*. Retrieved from https://www.hamburgwasser.de/umwelt/energiegewinnung/biogas.

SA Water. *Overview: Nais—the Future of Food Production in Sa*. Retrieved from https://www.sawater.com.au/nais/overview.

SFPUC. *Onsite Water Reuse Grant*. Retrieved from https://www.sfpuc.gov/programs/grants/onsite-water-reuse-grant#:~:text=Our%20Onsite%20Water%20Reuse%20Grant,for%20non%2Dpotable%20uses%20such.

6

Water-Energy Consumption and Sustainable Behavior Practices

Yulia Ermolaeva

Abstract Water-Energy Consumption and Sustainable Behavior Practices: This chapter explores how theories such as Norm Activation Theory (NAT), Value-Belief-Norm (VBN), Theory of Planned Behavior (TPB), and behavioral economics inform interventions to reduce water and energy use with practical applications and highlights in interplay of objective policy, technology, and socio-economic factors in shaping subject practices household resource use, underscoring the need for regionally tailored strategies to achieve sustainability goals with cross-cultural observation of the social practices and mechanisms of social policy in energy/water consumption and water/energy saving activities.

Keywords Water · Energy consumption in households · Environmental sociology · Environmental psychology · Subject approach · Norm Activation Theory (NAT) · Value-Belief-Norm (VBN) · Theory of Planned Behavior (TPB) · And behavioral economics

Introduction

The nexus between water and energy consumption is a critical sustainability challenge, exacerbated by climate change and population growth. Addressing this requires understanding human behavior through social science frameworks its needs of approaches from social sciences, conceptualizing and measuring structural and subjective aspects of social transformations in water and energy consumptions in social sciences.

Social Sciences Theories in Study of Energy and Water Saving in Households

Values-Beliefs-Norms

The psychological underpinnings of resource conservation in households have been extensively explored through the lens of Values-Identity-Personal norms (VIP) theory and Values-Beliefs-Norms (VBN) theory. VBN theory integrates values, beliefs, and norms to explain environmental action (Stern, 2000). Individuals with biospheric values perceive threats to ecosystems (ecological worldview), feel obligation (personal norm), and act accordingly. T. Dietz (Stern, P.C.; Kalof, L.; Dietz, T.; Guagnano, G.A., 1995) highlighted the role of social norms in promoting low-flow fixtures, while Larson et al. (2009) linked VBN to community-led water audits in Australia, cutting municipal use by 15%. Both frameworks posit that biospheric values (concern for environmental sustainability) initiate a causal chain leading to conservation behaviors. Biospheric values strengthen an individual's environmental self-identity—the cognitive categorization of oneself as "environmentally conscious"—which subsequently activates personal norms (feelings of moral obligation) to conserve resources. A nationally representative U.S. study confirmed that environmental self-identity mediated the relationship between biospheric values and conservation behaviors across food, energy, and water domains. Notably, hedonic values (prioritization of pleasure/comfort) and egoistic values (economic self-interest) negatively correlated with conservation actions, revealing motivational conflicts (Fuenfschilling & Truffer, 2016). This aligns with Maslow's hierarchy, where safety/security needs (Level 2) often compete with self-actualization (Level 5) when conservation requires effortful behavior change.

Norm Activation Theory (NAT)

NAT posits that pro-environmental behavior arises when individuals recognize adverse consequences (awareness of consequences, AC), feel responsible (ascription of responsibility, AR), and believe in their efficacy to mitigate harm (personal norm, PN) (Schwartz, 2006). Applied to water conservation, studies demonstrate that campaigns emphasizing drought impacts (AC) and personal accountability (AR) effectively reduce consumption. For instance foundings that Dutch households exposed to messages linking water waste to global scarcity reported stronger PN and adopted water-saving technologies ("The Norm Activation Model and Theory-Broadening: Individuals' Decision-Making on Environmentally-Responsible Convention Attendance," 2014). Similarly, Bamberg and Möser showed that AR-mediated interventions in Germany increased rainwater harvesting (Moser et al., 2022).

Sociocultural and Contextual Mediators

The social context critically shapes conservation motivation through normative influence and family socialization. Social norms operate via two primary mechanisms: the norm of social responsibility (helping those dependent on us) and the norm of reciprocity (expecting mutual aid). These norms, however, are vulnerable to diffusion of responsibility in collective settings, where individuals assume others will intervene—particularly when conservation costs are high. In family contexts, parental modeling and shared practices instill enduring habits (Kim & Kim, 2024). Children develop conservation-oriented identities through routines like regulated screen time, participation in outdoor activities, and family discussions about utility bills. Such practices foster intrinsic motivation when framed as value-driven choices rather than imposed restrictions. Conversely, purely extrinsic motivators (financial penalties) may undermine intrinsic commitment if perceived as controlling rather than informational.

Theory of Planned Behavior (TPB)

TPB asserts that intention—shaped by attitudes, subjective norms, and perceived behavioral control (PBC) predicts behavior (Ajzen, 1991) demonstrated that positive attitudes toward conservation and peer influence (subjective norms) drove water-saving. Research according PBC approach in Australia shown providing low-cost shower timers in Melbourne increased perceived efficacy, reducing consumption by 20%. Yuriev et al. (2018) reinforced this, showing that subsidies for efficient appliances enhanced PBC in Canadian households.

Behavioral Economics

Behavioral economics challenges rational choice models, emphasizing cognitive biases and heuristics. Thaler and Sunstein's (2008), "nudges" have been widely applied. Physical-technical systems constrain or enable conservation by mediating agency. Water consumption exemplifies a "random and unmanaged process" influenced by:

Technical factors: pressure losses, inefficient fixtures, and hot water temperatures that increase "non-useful" consumption.
Temporal factors: aging infrastructure causing leaks that escalate with system pressure and pipe degradation.
Economic factors: property characteristics (e.g., pool ownership) and appliance efficiency, which are often income-dependent.

Based on empirical research, several aspects of household water conservation present significant challenges for individuals, primarily stemming from the invisibility of water flows and consumption, infrastructural constraints, entrenched habits, perceived trade-offs with comfort/

hygiene, and split incentives. The lack of real-time feedback on usage makes it difficult for residents to monitor consumption or detect leaks effectively, reducing perceived control and agency. Aging infrastructure, including inefficient fixtures, hidden leaks, and high system pressure, contributes substantially to "non-useful" consumption, often beyond the occupant's direct awareness or immediate capacity to rectify (Fielding et al., 2013). Modifying ingrained, automatic behaviors associated with personal hygiene (for example, long showers) and appliance use (running dishwashers half-full) is particularly resistant to change due to habit strength and comfort associations. Furthermore, reducing water use for activities perceived as essential to well-being or cleanliness (gardening, laundry frequency) often triggers psychological reactance or perceived sacrifices (Willis et al., 2011). In rental properties, the landlord-tenant problem creates a major barrier, where tenants lack incentive to invest in efficiency upgrades (low-flow fixtures) while landlords avoid costs that don't directly reduce their bills. These challenges are compounded by the perception of water as an abundant "public good" rather than a scarce resource requiring individual stewardship (Millock & Nauges, 2010). These structural barriers interact with psychological ownership: When individuals perceive water as a "gift of nature" rather than a "commodity" requiring costly processing, conservation motivation diminishes. Metering with volumetric billing counteracts this by creating economic feedback loops, linking behavior to personal costs. In summary, Household water-use behavior is associated with three necessary conditions: capability, opportunity, and motivation (Addo et al., 2018).

Water and Energy Consumption in Households: Global Statistics and Regional Challenges

The balance between household water and energy consumption and conservation efforts varies significantly across regions, influenced by economic development, policy frameworks, and cultural norms. While high-income countries often exhibit higher per capita consumption due to advanced infrastructure and lifestyle demands, they also lead in implementing efficiency technologies. Conversely, low- and middle-income countries face challenges such as water scarcity, energy poverty, and limited access to sustainable technologies, despite lower per capita use. Climate change exacerbates these disparities, with arid regions and developing economies disproportionately affected by resource stress. Below, comparative statistics and case studies illustrate these dynamics, followed by an analysis of regional leaders and challenges (Table 6.1).

Table 6.1 Household water consumption and savings (selected regions). (*Statistics | UN World Water Development Report, 2025*)

Region	Average yearly water use (m³) for region approximently	Average daily water use per capita (litres) for region	Key savings practices
Africa	12,860,000,000	549	Crisis pricing, public awareness campaigns
Latin America	6,728,000,000	1,033	
Northern Europe	2,472,000,000	1,068	Rainwater harvesting, tariff incentives
Southern Europe	5,398,000,000	1,289	
BRICS	226,500,000,000	1,703	Community-led rainwater storage, Efficient irrigation subsidies
Middle East	7,316,000,000	2,014	Desalination plant investments
United States	444,300,000,000	3,732	Low-flow fixtures, smart meters
South Asia	168,400,000,000	1,975	
Australia and New Zealand	21,330,000,000	2,046	Mandatory water restrictions, rebates

Water consumption patterns and water-saving technologies vary significantly across regions due to differences in climate, economic development, population density, and water availability (*Water Consumption Statistics—Irrigreen*, 2025). In arid regions such as the Middle East, where high temperatures and low rainfall reduce water availability, countries like Saudi Arabia and the United Arab Emirates rely heavily on desalination and implement strict water conservation policies, including subsidies for efficient irrigation and use of treated wastewater in agriculture. In contrast, South Asia, with its monsoon-driven climate and large rural population, faces challenges related to seasonal water scarcity and poor infrastructure, prompting the adoption of decentralized solutions such as rainwater harvesting, micro-irrigation, and improved cookstoves that also reduce water-linked energy use (International Statistics of Water Services, 2014). Latin America, despite relatively abundant freshwater resources, experiences regional disparities due to deforestation, urbanization, and uneven distribution, leading to increased use of smart metering, leak detection systems, and community-based demand management programs (Humanities et al., 1999). North and South Europe demonstrate advanced water-saving practices rooted in regulatory frameworks and technological innovation, including greywater recycling, low-flow fixtures, and stringent industrial water reuse standards, driven by both environmental awareness and high water tariffs (Zhu et al., 2024). In Africa, where per capita water consumption is among the lowest globally, access remains a critical issue, with many communities relying on hand pumps, solar-powered boreholes, and point-of-use filtration systems to improve efficiency and safety. Australia and New Zealand, facing increasing droughts due to climate change, have implemented integrated water resource (*GCC Statistical Center—Water Statistics in GCC Countries*, 2025) (Table 6.2).

Household energy consumption varies widely across regions due to disparities in income levels, climate conditions, access to modern energy services, and policy frameworks promoting energy efficiency (*Primary Energy Consumption by Region 2023| Statista*, 2023). In high-income regions such as North Europe, North America, and parts of Oceania, per capita household energy use is among the highest globally, driven

Table 6.2 Household energy consumption and savings (selected countries). Adapted from *World Power Consumption | Electricity Consumption | Enerdata* (2025), Ritchie (2021)

Africa	46,750,897,070,000	8.03%	34,541	Limited access; use of biomass, solar lanterns, and off-grid systems
Latin America	52,822,489,510,000	7.47%	65,769	Use of efficient appliances, solar water heaters, LED lighting, and time-of-use electricity tariffs
North Europe	126,247,224,800,000	19.45%	205,787	High-efficiency buildings, smart meters, heat pumps, insulation, renewable
South Europe	49,986,446,940,000	7.94%	133,969	Solar PV, energy-efficient appliances, passive design, and behavioral savings
BRICS	302,225,790,700,000	19.06%	133,092	Mixed: coal dependence in industry; growing adoption of solar, biogas, and appliance efficiency
Middle East	39,819,677,880,000	6.03%	225,387	Air-conditioning efficiency, solar rooftops, smart grids, and government subsidies for energy-saving appliances

United States	97,661,161,460,000	16.77%	293,979	Smart thermostats, ENERGY STAR appliances, home automation, insulation, and solar
South Asia	32,616,447,740,000	5.25%	29,456	Improved cookstoves, solar lights, microgrids, and behavior-based conservation
Australia and New Zealand	15,152,560,940,000	1.91%	237,678	Solar PV, efficient building codes, water heater timers, and public awareness campaigns

by widespread reliance on electric heating, cooling systems, and energy-intensive appliances (*Global Energy Demand by Region to 2050—Thunder Said Energy*, 2025). These regions have implemented advanced energy-saving measures including smart metering, building insulation standards (passive house design), and appliance labeling programs (such as ENERGY STAR), which aim to reduce demand while maintaining comfort (*Primary Energy Consumption by World Region*, 2025). For example, countries like Germany, Sweden, and the UK demonstrate relatively low per capita growth in household energy use despite high living standards, reflecting successful decoupling through technological improvements and behavioral shifts. In contrast, sub-Saharan Africa and South Asia exhibit the lowest per capita household energy consumption, largely due to limited access to electricity and reliance on traditional biomass for cooking and heating. Countries like Ethiopia, Malawi, and Afghanistan report some of the lowest energy use per capita, with many households relying on inefficient stoves that pose health risks and contribute to deforestation (*Ratio of Household Electricity Use to National Electricity Consumption by Country | Center For Global Development*, 2025). However, recent initiatives focusing on off-grid solar solutions, improved cookstoves, and microgrids are gradually increasing access while reducing environmental and health impacts.

Despite these efforts, energy poverty remains a significant barrier to development in these regions, where electrification rates often fall below 50% in rural areas. Latin America and parts of Southeast Asia show intermediate levels of household energy consumption, with growing demand driven by urbanization and rising incomes. Countries like Brazil, Indonesia, and Mexico are experiencing increased adoption of air conditioning, refrigeration, and electronic devices, leading to upward pressure on residential energy use (*Net Zero by 2050—Analysis—IEA*, 2025). At the same time, several governments in this group are promoting demand-side management programs, subsidies for efficient appliances, and time-of-use pricing to curb peak loads and encourage conservation.

The Middle East and Gulf states, particularly Qatar, Kuwait, and the UAE, display extremely high per capita household energy consumption, primarily due to intensive cooling demands and historically low energy prices 2025. While these countries consume large amounts of energy at the household level, recent investments in solar PV installations, smart grid technologies, and mandatory energy performance codes for buildings are beginning to moderate consumption growth. Similarly, Australia and New Zealand, facing increasing climate pressures, have strengthened regulatory frameworks around home energy ratings and renewable integration to address rising residential demand (*Average Household Support for Policy Measures & Perceived Effectiveness, 2022—Charts—Data & Statistics—IEA*, 2025). Overall, household energy consumption patterns reflect a complex interplay between geographic, economic, and policy factors, with divergent trajectories across regions. While industrialized nations focus on deep decarbonization and efficiency optimization, developing economies grapple with expanding access and minimizing the health and environmental costs of traditional energy use. Bridging these gaps requires regionally tailored policies, targeted investment in clean technologies, and sustained international cooperation.

Case Studies: Implementation Theories into Practice

Leveraging Behavioral Economics for Sustainable Water and Energy Consumption

Behavioral economics, which integrates psychological insights into economic decision-making, offers innovative tools to address the intention-action gap in sustainable consumption. By targeting cognitive biases (present bias, status quo preference, and social norm adherence, which most common in resource saving problems) these strategies nudge individuals and institutions toward resource-efficient behaviors without coercive mandates (Tables 6.3, 6.4).

Behavioral economics enhances environmental sustainability efforts by addressing psychological and decision-making barriers that traditional policies often overlook, offering innovative mechanisms to drive conservation. This approach deeply studying leveraging cognitive biases such as present bias, social conformity, and status quo bias, behavioural

Table 6.3 Behavioral economics interventions in water savings. Booysen et al. (2019), Addo et al. (2018), Humanities et al. (1999), Dolnicar et al. (2012)

Principle	Intervention	Case study	Outcome
Social norms	Comparative billing (peer consumption feedback)	Cape Town, South Africa (2018)	50% reduction in household use during "Day Zero" crisis
Default effect	Pre-installed low-flow fixtures	Melbourne, Australia (2000s drought)	35% drop in per capita use
Loss aversion	Tiered pricing penalizing excess use	Tucson, Arizona, USA	20% reduction in peak demand
Prosocial incentives	Public recognition for "Water Champions"	Valencia, Spain (community irrigation)	15% decline in agricultural misuse
Framing	Emotive messaging on drought impacts	São Paulo, Brazil (2014–2016 crisis)	25% voluntary cuts

Table 6.4 Behavioral economics interventions in energy savings. Allcott (2011), Kinoshita (2020), Pichert and Katsikopoulos (2008)

Principle	Intervention	Case Study	Outcome
Social Comparisons	Home energy reports with neighbour rankings	Opower Program (USA, EU, Australia)	2–5% average household reduction
Commitment Devices	Pledges to reduce usage paired with feedback	Kyoto, Japan (Cool Biz Campaign)	11% drop in office AC consumption
Scarcity Messaging	Real-time grid stress alerts	California, USA (Flex Alerts, 2020–2023)	9% peak demand reduction
Gamification	App-based rewards for energy-saving milestones	Bangalore, India (Smartron)	12% decrease in residential use
Default Green Energy	Opt-out renewable energy tariffs	Germany (Green Electricity Tariff)	80% enrolment in green energy

tools like commitment devices that help individuals align immediate actions with long-term environmental goals. Similarly, social norm feed-back—demonstrated through the Opower program—has proven more effective than financial incentives in reducing energy consumption by tapping into people's tendency to conform to peer behavior. Defaults in green energy enrollment, as implemented in Germany, exploit status quo bias to increase renewable adoption without limiting choice, thereby simplifying pro-environmental decisions. These strategies also contribute to broader sustainability discourse by bridging equity gaps, as seen in Nairobi slums where prepaid systems improved access while reducing overuse. Furthermore, they enhance policy flexibility by adapting to diverse cultural contexts, exemplified, which successfully reframed energy reduction as a collective responsibility scalability and cost-efficiency are evident in the global replication. Thus, hybrid frameworks in behavioral economics combining nudges with **equitable pricing** and **technology access** are critical for sustained impact, reframes sustainability as a series of manageable, socially reinforced choices, offering a pragmatic pathway to reduce water and energy footprints. By aligning interventions with human psychology, policymakers can amplify the impact of technical and regulatory measures, fostering a culture of conservation integral to global environmental resilience.

Successful Environmental Practices Driven by Sustainable Behavior Approaches

Sustainable behavior approaches have catalyzed transformative environmental practices worldwide, demonstrating how psychological insights and socio-cultural strategies can drive resource efficiency. Below presented global case studies that highlight the application of behavioral theories—such as social norms, nudges, and value-based interventions—to reduce water and energy consumption, emphasizing measurable outcomes and scalability (Tables 6.5, 6.6).

Social norms play a pivotal role in shaping sustainable behavior by influencing individual choices through collective expectations and peer

Table 6.5 Case studies in water conservation. Best Practices for Driving Environmentally Sustainable Behavior Change (2025), Roberts (2019), Klaniecki et al. (2018)

Country	Behavioural principle	Strategy	Outcome
Cape Town, South Africa	Loss Aversion & Crisis Framing	Tiered tariffs, "Day Zero" public alerts	50% household reduction (2018 crisis)
Melbourne, Australia	Theory of Planned Behaviour (TPB)	Mandatory restrictions+rebate programs	40% drop in per capita use (2000s drought)
Rajasthan, India	Community-Based Social Norms	Participatory groundwater management	30% rise in aquifer levels (2010–2020)
Valencia, Spain	Prosocial Incentives	Public recognition for efficient farmers	20% agricultural water savings
Singapore	Value-Belief-Norm (VBN) Theory	National water education campaign	40% household reduction since 2000

Table 6.6 Case studies in energy conservation. *Encouraging Pro-Environmental Behaviour: An Integrative Review and Research Agenda—ScienceDirect* (2025), Adegoke et al. (2023)

Country	Behavioural principle	Strategy	Outcome
United States	Social Norm Comparisons	Opower's Home Energy Reports	2% avg. household reduction (12 million homes)
Japan	Social Proof & Commitment	Cool Biz Campaign (casual office attire)	1.4 million tons CO_2 saved annually
Rwanda	Community-Based Collective Action	Green Village eco-communities	60% adoption of solar stoves (2015–2023)
Germany	Default Effect	Opt-out green energy tariffs	80% enrolment in renewables
Kenya	Commitment Devices	Prepaid solar meter systems	50% rural energy access increase

comparisons. Programs like Opower's energy feedback system demonstrate how providing households with information about their energy use relative to their neighbors can effectively reduce consumption, as individuals are motivated to conform to perceived standards of efficiency. This normative influence is further amplified when framed within the context of potential loss, messaging emphasizing the imminent depletion of water supplies triggered immediate behavioral adjustments across some cities. Similarly, community-led initiatives benefit from the power of social norms by embedding sustainability into shared governance structures. In Rajasthan, India, local farmers collectively managed groundwater resources, leading to the revival of dried wells and reinforcing cooperative conservation practices that align with established behavioral theories of collective action. These examples illustrate how norms operate not only at the individual level but also through community dynamics that reinforce pro-environmental behaviors.

Cultural context significantly shapes the effectiveness of norm-based interventions,, which reframed energy reduction as a societal duty rather than a personal sacrifice. By aligning policy goals with cultural values of conformity and collective responsibility, the initiative successfully reduced reliance on air conditioning without mandates or financial incentives. This approach highlights the importance of tailoring behavioral strategies to fit local norms and values to ensure relevance and acceptance. Structural nudges, such as Germany's opt-out green energy tariffs, further exploit habitual decision-making by setting sustainability as the default option, thereby increasing adoption rates without limiting consumer freedom. Such interventions reflect the broader principle that subtle changes in choice architecture can yield significant environmental benefits when aligned with psychological tendencies like inertia and social conformity.

Equity considerations are also central to the success of norm-driven sustainability efforts. Kenya's prepaid solar meter model, for instance, leveraged commitment devices to enable low-income households to transition away from polluting fuels while maintaining affordability. This approach not only improved access but also reinforced responsible usage patterns by integrating behavioral insights with financial inclusivity. Despite these successes, challenges remain in sustaining long-term engagement, particularly once immediate threats subside, as observed in post-drought water consumption trends in Melbourne. Additionally, cultural mismatches in intervention design can hinder scalability, underscoring the need for localized adaptation, as exemplified by Rwanda's Green Villages initiative, which integrated sustainability measures into traditional practices to ensure lasting impact. Collectively, these insights highlight the transformative potential of social norms in driving sustainable behavior, provided that interventions are culturally sensitive, equitably designed, and supported by enduring structural mechanisms.

Modern Psychological Approach in Promotion of Energy and Water Saving Behavior

Modern psychology offers a range of evidence-based methodologies that can be effectively applied to influence public behavior and decision-making processes. These approaches are grounded in cognitive and behavioral theories that emphasize the interplay between individual cognition, environmental factors, and social contexts.

Cognitive-Behavioral Therapy (CBT). CBT focuses on identifying and modifying maladaptive thought patterns that may hinder pro-environmental behaviors focuses on the relationship between a person's thoughts, feelings, and behavior (Beck & Beck, 2020). The main goal of CBT is to teach a person to identify and change maladaptive thoughts and behavior that affect their emotional state and quality of life in terms of sustainable behavior. Tools commonly used within this framework include cognitive restructuring, behavioral activation, exposure techniques, self-monitoring, problem-solving training, goal setting, and thought records (Bentler et al., 2023). Research demonstrates that CBT interventions can significantly improve individuals' willingness to adopt resource-saving habits by altering negative beliefs and reinforcing positive behavioral changes (Homburg & Stolberg, 2006).

Acceptance and Commitment Therapy (ACT). ACT encourages individuals to accept their thoughts and feelings rather than resist them, while committing to actions aligned with personal values. Key tools include mindfulness exercises, values clarification, cognitive defusion, acceptance, committed action, present-moment awareness, and self-as-context (Hayes & Strosahl, 2005). The effectiveness of ACT in helping individuals align decisions with long-term sustainability goals, even when faced with short-term discomfort or economic concerns was approved by empirical studies about climate distress and creation of sustainable household practices (Williams & Samuel, 2024).

Motivational Interviewing (MI). MI is a client-centered counseling style designed to enhance intrinsic motivation for change by exploring and resolving ambivalence (Römer et al., 2024). Techniques include open-ended questions, reflective listening, affirmations, summarizing, developing discrepancy, change talk elicitation, and rolling with resistance, offering empirical validation of MI's success in fostering pro-environmental behavior through increased personal commitment (Klonek et al., 2015).

Social Cognitive Theory (SCT). SCT provides a framework for understanding how individuals learn from observing others within social contexts. Tools include observational learning, modeling, outcome expectations, reinforcement, goal setting, self-efficacy scales, and vicarious learning. Bandura (1986) introduced the theory, which has since been applied in numerous environmental campaigns. Researchers demonstrated how modeling and self-efficacy strategies can foster widespread adoption of conservation practices within communities (Sawitri et al., 2015).

Contingency Management (CM). CM increases desirable behaviors through systems of reinforcement and punishment. Tools include incentive-based feedback, reward systems, token economies, penalty structures, usage tracking, behavioral contracts, and positive reinforcement schedules. Furlan Matos Alves et al. (2017) Social research showed that financial incentives and structured feedback mechanisms can significantly alter consumption patterns and promote sustainable use of resources (Furlan Matos Alves et al., 2017).

Modern psychology and specialized approaches in social sciences unite subjects for contribute to effective social practices by informing macro-level policies that address socio-environmental challenges through behavioral insights, these approaches support institutions and organizations at meso level in making decisions that align with sustainable development goals and foster pro-environmental behavior within communities. At the micro level, psychological interventions help shape household practices by promoting energy and water conservation through tailored cognitive-behavioral strategies.

References

Addo, I. B., Thoms, M. C., & Parsons, M. (2018). Household water use and conservation behavior: A meta-analysis. *Water Resources Research, 54*(10), 8381–8400. https://doi.org/10.1029/2018WR023306

Adegoke, M., Alaka, H., Ajayi, S., & Popoola, M. O. (2023). *Optimizing energy efficiency and sustainability: Building energy consumption prediction with broad learning system and alternating direction method of multipliers.* https://doi.org/10.2139/ssrn.4620845

Ajzen, I. (1991). The theory of planned behavior. *Theories of Cognitive Self-Regulation, 50*(2), 179–211. https://doi.org/10.1016/0749-5978(91)900 20-T

Allcott, H. (2011). Social norms and energy conservation. *Special Issue: The Role of Firms in Tax Systems, 95*(9), 1082–1095. https://doi.org/10.1016/j.jpubeco.2011.03.003

Average household support for policy measures and perceived effectiveness, 2022—Charts—Data & Statistics—IEA. (2025). Retrieved June 6, 2025, from https://www.iea.org/data-and-statistics/charts/average-household-support-for-policy-measures-and-perceived-effectiveness-2022.

Bandura, A. (1986). *Social foundations of thought and action: A social cognitive theory.* Prentice Hall.

Beck, J. S., & Beck, A. T. (Foreword). (2020). *Cognitive behavior therapy: Basics and beyond* (3rd Ed.). Guilford Press.

Bentler, D., Kadi, G., & Maier, G. W. (2023). Increasing pro-environmental behavior in the home and work contexts through cognitive dissonance and autonomy. *Frontiers in Psychology, 14,* 1199363. https://doi.org/10.3389/fpsyg.2023.1199363

Best Practices for Driving Environmentally Sustainable Behavior Change. (2025). Food For Climate League. Retrieved June 6, 2025, from https://www.foodforclimateleague.org/store/p/best-practices-environmentally-sustainable-behavior-change.

Booysen, M. J., Visser, M., & Burger, R. (2019). Temporal case study of household behavioural response to Cape Town's "Day Zero" using smart meter data. *Water Research, 149*, 414–420. https://doi.org/10.1016/j.watres.2018.11.035

Dolnicar, S., Hurlimann, A., & Grün, B. (2012). Water conservation behavior in Australia. *Journal of Environmental Management, 105*, 44–52. https://doi.org/10.1016/j.jenvman.2012.03.042

Encouraging pro-environmental behaviour: An integrative review and research agenda—ScienceDirect. (n.d.). Retrieved April 10, 2023, from https://www.sciencedirect.com/science/article/abs/pii/S0272494408000959?via%3Dihub.

Fielding, K. S., Spinks, A., Russell, S., McCrea, R., Stewart, R., & Gardner, J. (2013). An experimental test of voluntary strategies to promote urban water demand management. *Journal of Environmental Management, 114*, 343–351. https://doi.org/10.1016/j.jenvman.2012.10.027

Fuenfschilling, L., & Truffer, B. (2016). The interplay of institutions, actors and technologies in socio-technical systems—An analysis of transformations in the Australian urban water sector. *Technological Forecasting and Social Change, 103*, 298–312. https://doi.org/10.1016/j.techfore.2015.11.023

Furlan Matos Alves, M. W., Lopes de Sousa Jabbour, A. B., Kannan, D., & Chiappetta Jabbour, C. J. (2017). Contingency theory, climate change, and low-carbon operations management. *Supply Chain Management, 22*(3), 223–236. https://doi.org/10.1108/SCM-09-2016-0311

GCC Statistical Center—Water Statistics in GCC Countries. (n.d.). Retrieved June 6, 2025, from https://gccstat.org/en/statistic/publications/water-statistics-bulletin-in-gcc-countries.

Global energy demand by region to 2050—Thunder Said Energy. (n.d.). Retrieved June 6, 2025, from https://thundersaidenergy.com/downloads/global-energy-demand-by-region-and-through-2050/.

Hayes, S. C., & Strosahl, K. D. (Eds.). (2005). *A practical guide to acceptance and commitment therapy.* Springer.

Homburg, A., & Stolberg, A. (2006). Explaining pro-environmental behavior with a cognitive theory of stress. *Journal of Environmental Psychology, 26*(1), 1–14. https://doi.org/10.1016/j.jenvp.2006.03.003

Humanities, I. A. of S. and, Technology, P. A. for S. and, Society, R. S., & East, J. and N. A. of S. (US) C. on S. W. S. in the M. (1999). Factors affecting patterns of water use. In *Water for the future: The West Bank and Gaza Strip, Israel, and Jordan.* National Academies Press (US). Retrieved from https://www.ncbi.nlm.nih.gov/books/NBK230258/.

International statistics of water services. (2014). The international water association.

Kim, S., & Kim, S. (2024). Does social value matter in energy saving behaviors?: Specifying the role of eleven human values on energy saving behaviors and the implications for energy demand policy. *Energy Strategy Reviews, 52,* Article 101327. https://doi.org/10.1016/j.esr.2024.101327

Kinoshita, S. (2020). Conjoint analysis of Japanese households' energy-saving behavior after the earthquake. *Energy & Environment, 31*(4), 676–691. JSTOR.

Klaniecki, K., Wuropulos, K., & Hager, C. (2018). *Behaviour change for Sustainable Development.* https://doi.org/10.1007/978-3-319-63951-2_161-1

Klonek, F. E., Güntner, A. V., Lehmann-Willenbrock, N., & Kauffeld, S. (2015). Using Motivational Interviewing to reduce threats in conversations about environmental behavior. *Frontiers in Psychology, 6,* 1015. https://doi.org/10.3389/fpsyg.2015.01015

Larson, K. L., White, D. D., Gober, P., Harlan, S., & Wutich, A. (2009). Divergent perspectives on water resource sustainability in a public–policy–science context. *Environmental Science & Policy, 12*(7), 1012–1023. https://doi.org/10.1016/j.envsci.2009.07.012

Middle East—Countries & Regions—IEA. (2025). Retrieved June 6, 2025, from https://www.iea.org/regions/middle-east/electricity

Millock, K., & Nauges, C. (2010). Household adoption of water-efficient equipment: The role of socio-economic factors, environmental attitudes and policy. *Environmental and Resource Economics, 46*(4), 539–565. https://doi.org/10.1007/s10640-010-9360-y

Moser, D., Steiglechner, P., & Schlueter, A. (2022). Facing global environmental change: The role of culturally embedded cognitive biases. *Environmental Development, 44,* Article 100735. https://doi.org/10.1016/j.envdev.2022.100735

Net Zero by 2050—Analysis—IEA. (2024). Retrieved February 12, 2024, from https://www.iea.org/reports/net-zero-by-2050.

Pichert, D., & Katsikopoulos, K. (2008). Green Defaults: Information Presentation and Pro-environmental Behaviour. *Journal of Environmental Psychology, 28,* 63–73. https://doi.org/10.1016/j.jenvp.2007.09.004

Primary energy consumption by region 2023| Statista. (2023). Retrieved June 6, 2025, from https://www.statista.com/statistics/263457/primary-energy-consumption-by-region/.

Primary energy consumption by world region. (n.d.). Our World in Data. Retrieved June 6, 2025, from https://ourworldindata.org/grapher/primary-energy-consumption-by-region.

Ratio of Household Electricity Use to National Electricity Consumption by Country | Center For Global Development. (2025). Retrieved June 6, 2025, from https://www.cgdev.org/media/ratio-household-electricity-use-national-electricity-consumption-country.

Ritchie, H. (2021). Global comparison: How much energy do people consume? *Our World in Data.*

Roberts, E. (2019). *Sustainable behaviour and environmental practices* (pp. 161–170). https://doi.org/10.4324/9781315179780-17

Römer, J., Herrmann, A., Molkentin, K., & Müller, B. S. (2024). Application of Motivational Interviewing in climate-sensitive health counselling—A workshop report. *Zeitschrift Für Evidenz, Fortbildung und Qualität Im Gesundheitswesen, 189*, 50–54. https://doi.org/10.1016/j.zefq.2024.07.003

Sawitri, D. R., Hadiyanto, H., & Hadi, S. P. (2015). Pro-environmental behavior from a Social Cognitive Theory Perspective. *Procedia Environmental Sciences, 23*, 27–33. https://doi.org/10.1016/j.proenv.2015.01.005

Schwartz, S. H. (2006). A theory of cultural value orientations: Explication and applications. *Comparative Sociology, 5*, 137–182.

Statistics | UN World Water Development Report. (2025). Retrieved June 6, 2025, from https://www.unesco.org/reports/wwdr/en/2024/s

Stern, P. C. (2000). New environmental theories: Toward a coherent theory of environmentally significant behavior. *Journal of Social Issues, 56* (3), 407–424. https://doi.org/10.1111/0022-4537.00175

Stern, P. C., Kalof, L., Dietz, T., & Guagnano, G. A. (1995). Values, beliefs, and proenvironmental action: Attitude formation toward emergent attitude objects. *V, 25*, 1611–1636.

Thaler, R., & Sunstein, C. (2008). *Nudge: Improving decisions about health, wealth, and happiness.* Yale University Press.

The norm activation model and theory-broadening: Individuals' decision-making on environmentally-responsible convention attendance. (2014). *Journal of Environmental Psychology, 40*, 462–471.https://doi.org/10.1016/j.jenvp.2014.10.006

Water Consumption Statistics—Irrigreen. (2025). Retrieved June 6, 2025, from https://irrigreen.com/pages/water-consumption-statistics?srsltid=AfmBOoqpL7Ql8YoX7iqUvqg7VOe2mV1YijWNOp-sdeUq_Je-LHCfUj04

Williams, M. O., & Samuel, V. M. (2024). Acceptance and commitment therapy as an approach for working with climate distress. *The Cognitive*

Behaviour Therapist, 17, Article e35. https://doi.org/10.1017/S1754470X 23000247

Willis, R., Stewart, R., Panuwatwanich, K., Williams, P., & Hollingsworth, A. (2011). Quantifying the influence of environmental and water conservation attitudes on household end use water consumption. *Journal of Environmental Management, 92*, 1996–2009. https://doi.org/10.1016/j.jenvman. 2011.03.023

World Power consumption | Electricity consumption | Enerdata. (2025). Retrieved June 6, 2025, from https://yearbook.enerdata.net/electricity/electricity-dom estic-consumption-data.html.

Yuriev, A., Boiral, O., Francoeur, V., & Paillé, P. (2018). Overcoming the barriers to pro-environmental behaviors in the workplace: A systematic review. *Journal of Cleaner Production, 182*, 379–394. https://doi.org/10. 1016/j.jclepro.2018.02.041

Zhu, X., Hou, M., & Wei, J. (2024). Global water use and its changing patterns: Insights from OECD countries. *Water, 16*(24). https://doi.org/ 10.3390/w16243592

7

Exploring the Complex Interdependence Nature of Marine Renewable Energy Sector: A Developing Country Perspective

Obodai Torto and Pius Siakwah

Abstract This chapter explores the institutional and governance complexities shaping marine renewable energy (MRE) development in Ghana, a lower-middle-income country navigating global climate commitments and domestic energy deficits. Drawing on polycentric governance theory, it analyzes how fragmented institutional frameworks, unequal power dynamics among stakeholders, and macroeconomic challenges hinder Ghana's energy transition goals. Despite abundant renewable potential such as tidal and wave energy Ghana's reliance on fossil fuels persists due to weak regulatory systems, limited private-sector engagement, and inadequate technological capabilities. The study highlights the need for coherent policy coordination, investment in local capacity-building, and adaptive governance structures to address the interplay of social, economic, and political factors in MRE adoption. Examining Ghana's energy transition within the broader context of sub-Saharan Africa, the chapter underscores the critical role of institutional quality and inclusive stakeholder collaboration in achieving sustainable energy systems amid global poly-crisis trends.

© The Author(s), under exclusive license to Springer Nature
Switzerland AG 2025
Y. Ermolaeva et al., *Rethinking Water and Energy for a Sustainable Future*, Palgrave
Studies in Climate Resilient Societies, https://doi.org/10.1007/978-3-032-04485-3_7

Keywords Marine Renewable Energy · Polycentric governance · Energy transition · Ghana · Policy environment · Developing countries · Sustainability · Governance complexity · Capability building

Introduction

The current global system major concerns with climate change and sustainability anchored in the net-zero carbon emission and just energy transition policy mantra have reconfigured multiplicity of nascent climate action policies. Thus, through the Paris agreement, governments of 195 countries in 2015 pledged commitment to hold global warming below 2 degree Celsius above preindustrial levels and ultimately limit it to 1.5 degree Celsius (World Bank, 2024). Importantly, the energy variable is eminent in the climate change and sustainability discourse and practice. Additionally, the energy factor which is a crucial factor in growth and development of a country. Also, it is a major source of climate-related pollution of the ecosystem that would inevitably instigate primary government and international policy attention (World Bank, 2023, 2024). It is therefore no surprising that many countries across the planetary scales have designed diverse policy instruments and strategies to optimize alternative energy sources. Hence the new wave with the energy transition. On a global scale six countries, namely; China, the United States, the EU27, India, Russia and Japan in 2021 comprised the world's largest CO_2 emitters (EDGAR, 2022). Additionally, those countries represent 66.4% of global fossil fuel consumption and 67.8% of global fossil CO_2 emissions (BP, 2022).

There appears to be an uneven energy capacities akin to the growing disparities between the advanced economies and sub-Saharan African (SSA) states. Therefore, from continental standpoint, many African economies continue to experience significant shortage of energy capacity which accentuates the continent's industrialization. Notwithstanding this dismal energy condition, African governments zeal to attain the global energy transition target have initiated wide-ranging correlating energy policies, programs and strategies to that effect. From this vista, this chapter uses the Marine Renewable Energy (MRE) sector of Ghana as the case study to represent the developing or lower middle income country context.

This chapter seeks to explore the institutional and governance underpinnings of MRE. The driving rationale behind this relational focus is that energy generation is nested in multiple and possibly autonomous agencies and domains at different scales. In effect, it requires a multi-stakeholder effort. Undergirding this multi-stakeholder framework is the varying degrees of unequal power relations. These institutions given their innate differences in terms of mandate, resources, ideas, and influences are not expected to operate in linear terms, but rather in messy and complex ways. Also, it would be appropriate to tease out institutional and specific public policy priorities at the conceptual and discursive level. This is against the backdrop that numerous studies on MRE the have centered on economic efficiency, social and ecological dimensions and efficiency of technologies (Qui et al., 2019; Cui & Zhao, 2023). Furthermore, Ghana's national energy policy following the technical orientation identifies inadequate regulatory framework, low private sector participation, poor land ownership, inadequate data on renewable resources and their utilization as bottlenecks of the renewable energy (NEP, 2022). This chapter adopts a discursive methodological approach.

This chapter proceeds as follows: the first section discusses the theoretical perspective of polycentrism as the guiding framework. This will be followed by an overview of the renewable energy and Ghana energy situation. The next section deals with the MRE and the matching institutional and policy environment imperatives. Furthermore, the section will analyze some of the key foundational institutions necessary for a successful MRE development. The chapter ends with concluding remarks.

Concept of Polycentric Governance

The concept of polycentricity is prominent in public administration and policy bias social sciences denotes a pluralistic maze of decision-making processes of semi-autonomous entities to address a public good issue. It was first espoused by Ostrom et al. (1961), who argue that the lack or absence of single dominant leader in the pursuit of a local public good. In essence, there is a presence of diverse actors working at different sizes and independent of many others within the local sphere to realize the ultimate goal. Thus, they view it as many centers of decision-making that are formally independent of each other.

According to Garmestani and Benson (2013), polycentric systems depicts complex adaptive systems without a central authority controlling the processes and structures of the system. And that they are marked by multiple governance units at multiple scales, with each unit having some capacity to govern at its scale. Furthermore, polycentricity as posited by Stephan et al. (2019) in the wider sense point to processes, systems, and structures that produce and reproduce decision making. For instance, land management would empirically encapsulate various government agencies at different levels of complex processes of decision making, regulation, utilization. Importantly, polycentricity would entail processes of interdependence and contestations which are likely to manifest in the management of natural resources such as water, minerals, forests, and hydrocarbon.

From standard practice, energy from hydro, hydrocarbon or renewable marine energy particularly from a developing country perspective mimics this polycentricity as a result of bevy of institutional actors in the decision-making spaces. This is anchored in the meagre financial, weak technical capacity, poor infrastructure and import dependent nature of developing countries such as Ghana. Darko et al. (2019) and Lavers et al. (2024) studies of decision making processes of the hydro dams in Ghana and Ethiopia respectively show how decision making of dams elicits diverse array of actors. These different actors also evince multiple interests and manifest in indeterminate consequences. Indeed, these unpredictable processes could be linked to the Leftwich (2005) and Hickey (2013) position on politics of decision making as a commingle of negotiation, compromises and contestations associated with resource prioritization and allocation. Similar relations could be said between officials of state' institutions and external financiers, with the balance of power tilting in favor of external actors. It is therefore essential to understand the nature of ruling relations among these actors in the decision prioritization of the core components of energy generation. More importantly, is the need to understand why and how particular energy choice captured overriding policy attention unlike resettlement and environmental components (Hall & Branford, 2016). Again, the ideological and institutional culture of the diverse actors need to be understood in how they define, shape and implement core decisions.

It bears noting the critical role of the epistemic community, global financial and international development institutions as producers, translation, diffusion, and mutation of energy policymaking ideas. Thus, it is essential to understand the nature of the epistemic logic that frames benefits of particular energy source through modernization, development, and sustainability discourses. Furthermore, it is imperative to examine the assumptions that underpin the construction of meaning about the components of MRE. This reinforces our position that decision making and implementation are likely to be embedded in complex social relations, and not reducible to linear processes.

The failures associated with some of the energy production projects require renewed attention to understand the decisions that undergird them. From technical standpoint, such failures are reducible to capability and capital deficits. However, we contend that the problem is more complex and significantly rooted in the minuscule policy consideration for them. The shifting nature of the practices and priorities attest to the non-linear nature of the decision-making process. Again, the continuous energy demands in most developing countries that justify the need for further investment alternative energy needs thoughtful rethinking.

Significantly, understanding the dialectical decision-making processes, would be helpful to avoid the binary dichotomies such as macro–micro and local–global. Rather, it becomes imperative to expand the analytical gaze to understand the complexity of diversity of actors and their power relations. In effect, a complex understanding of the diverse actors would illuminate the messiness, contradictions, and complexity of processes that shape dam decision-making.

From an analytical standpoint, MRE production decision-making constitutes a critical object of analysis. We follow Gupta and Ferguson (1999) who posit the need to move away from thinking of field sites as geographical spaces, and view institutions and policy decision-making processes as sites for field research. Thus, by applying critical interrogation of renewable dam decision-making, there is the possibility to obtain nuanced understanding to impact on policy change and transformation.

Renewable Energy Context and Ghana Energy Situation

Globally, the drive for renewable energy is to increase energy mix due to production efficiency, transportation, distribution, and optimize end-use consumption. This quest for renewable energy requires the exploitation of better alternatives to the realization of the global and national energy transition goal. As a matter of fact, the UN Sustainable Development Goal (SDG) 7 is unambiguously clear on the need for countries to pursue affordable, reliable, sustainable and modern energy for all (Okereke, 2024; Cui & Zhao, 2023). Significantly, Ghana just like many developing countries being a signatory to the SDGs, has mainstreamed Goal 7 into its national energy policy toward poverty reduction and industrialization driven economic growth. Thus, the thrust of this renewable energy is to address the energy deficit against the backdrop of rising demands, promote clean energy technologies as well as promote international cooperation relative to energy matters. Indeed, since 2010, the various governments of Ghana have prioritized renewable energy in their policy blueprints. For instance, the National Energy Policy (NEP) (2021), has identified tidal wave, wind, solar and marine/ocean as major source of renewable energy.

Ghana has witnessed a significant rise in the national population from 25 million in 2010 to 30.8 million in 2021, which is projected to further grow by an annual average of 2% to 72.2 million in 2070. This is partly attributed to the current youthful demography of the country (GSS, 2021). Alongside the population growth is the growing or expanding urban population which has also increased the energy demands as the housing pressure and informal economic activities remain an upward slope. Furthermore, the National Energy Transition Framework (NETF) states that the urban–rural share is likely to rise from 56% in 2021 to 85% by 2070 resulting in an annual growth rate of 1% (NETF, 2023).

It is significant to note that Ghana has an informal labor market, which is due to lack of depth, diversity and linkages in the economy. Arguably, the youth bulge of Ghana's population in a way has accentuated the increase in the energy demands. Many of the youth are self-employed in non-tradable vocations such as hairdressing, barbering, and dressmaking. In effect, the dominant informal economy, youth bulge demography and the ever-present historical extractive sector have shaped the energy demands of the country.

More importantly, the transition from low developing country to lower-middle income country status in 2010 expectedly increased the national energy consumption. Thus, with the country's per capita income of $1225 and characterized by an electricity consumption of $364 kWh, witnessed an increment in both per capita income and energy consumption of $2337 and $546 kWh respectively (NEP, 2021). The logical implication is that as the country seeks to achieve higher economic growth and development notwithstanding the recent macroeconomic maelstroms, it is possible that the energy demands will continue to increase. Significantly, the country's annual electricity consumption rose from 347 kWh in 2010 to 546 kWh in 2020, is far below the 1000 kWh minimum global threshold for lower-income middle economies (NEP, 2021). Clearly, the country's quest to reach the upper middle income category via increased production is likely to require much energy generation and utilization. This high futuristic demand invariably reinforces the inevitable need to explore alternative sources as also contemporaneously tied to ratified global conventions and treaties. Hence, the renewable energy turn.

The issue of energy sources is very crucial to the renewable turn. Historically, the hydro has been the major source of Ghana's energy from 1965 to the late 1990s. However, the intermittent energy crises due to drought and other operational inefficiencies from the early 1980s led to shift to geo-thermal plants in 1997 to increase production. As of 2010, hydro contribution to electricity generation mix was 54.5% and oil accounted for 45.5%. However, year 2020 recorded a diminution of hydro's contribution to 29.9% and oil rose to 69.9% (NEP, 2021; NETP, 2022). Currently, oil is the dominant source of energy generation in Ghana to the detriment of renewable biomass. This dominance of oil in the wake of the net-zero emission turn has raised a sustainability quagmire. Indeed, Ghana as a signatory to the COP 26 commitments, to essentially phase down dependence on fossil fuel to reduce greenhouse gas emission reconfigured its energy sources. Moreover the Government of Ghana did prioritize renewable energy and reaffirmed the Paris Agreement plus other regional and international treaties and protocols. Therefore, the turn to renewable energy is a result of combination exogenous and endogenous factors.

The energy question has been a quintessential infrastructure resource of global, continental and national interest/priority, and more significant is the overwhelming choice for renewable energy. Therefore, it is not a coincidence for Ghana to anchor its energy strategy based on the Renewable Energy Act of 2011. The Act is grounded in its underlying premise: " meeting long-term demand via the use of public. Private, and international investment; acceleration of privatization processes; and to ensure optimal and sustainability in development and operating every source of renewable energy..." (Parliament of Ghana, Act 832.2011). Ghana's commitment to the realization of energy security and clean energy objective is grounded in the development of the National Energy Transition Framework (2022–2070). The preamble of this Framework states that:

> Ghana recognizes that the energy and transportation sectors are key areas in reducing emissions. Consequently, steps must be taken to transition these sectors towards a net-zero emissions future. To attain this, Ghana must transition to the production and utilization of clean energy and the implementation of measures to mitigate any emissions that occur in the process. This will ensure that Ghana contributes her quota to the reduction of global GHG emissions, and, more importantly, achieve decarbonization, energy access and security, and energy efficiency", (NETF, ii, 2023).

Following this framework is the specific core objectives of the Renewable Energy Master Plan (REMP). This framework is an essentially pro-investment orientated policy. Importantly, the REMP aims to achieve the following: (a) increase the proportion of renewable energy in the national energy mix; (b) reduce the dependence on biomass as the main fuel for thermal energy applications; (c) provide renewable energy-based decentralized electrification options in 1,000 off-grid communities; and (d) promote local content and local participation in the renewable energy industry.

The realization of these fine objectives mimics an elusive quest, which is reflective in many SSA countries pursuit of renewable energy. Agyekum (2020) identifies the inclement investment climate in African economies, such as high operational costs, poor infrastructure, and weak policy environment for the renewable energy under-performance.

Marine Renewable Energy Turn and Institutional Policy Environment Imperatives

Currently, China as the largest consumer of energy (EDGAR, 2022) has placed priority on the development of their vast MRE potential as found in the Bohai, Yellow, East China and South China seas (Qiu et al., 2019). A full development of China's MRE is likely to affect the country's industrial structure and ultimately enhance the energy mix.

The 550 km coastline of Ghana is deemed as having an immense potential for generating electricity. A wide array of renewable energy sources include; hydro, bioenergy, solar, and wind. Importantly, this tidal wave potential is currently under assessment to ostensibly augment the national energy mix. Further, the quest to optimize benefits in the blue economy has accentuated Ghana's interest in the MRE. From the outset, among the various renewable energy sources, that of marine is deemed to be of immense consistency and predictability (Qiu et al., 2019). MRE as evident in the name represents a novel antidote for coastal regions that offer decarbonized and scalable opportunities for energy security and resilience (Manasseh et al., 2017). Thus, MRE embodies technologies that mobilize energy potential from the ocean through waves, currents, tides, and salinity or thermal gradients and convert that potential into electricity or other usable forms of energy. Indeed, marine energy or blue energy is deemed to power remote communities and serve as a formidable complement to wind and solar generation. This plausible predictability renders particular value to grid balancing operators to guarantee a stable energy supply from renewable energy sources.

The Intergovernmental Panel on Climate Change (IPCC) reiterate technological bottlenecks to MRE, particularly tidal and wave. As a matter of fact, the technologies are quite novel even within the industrialized economies, and characterized by commercial uncertainties, investments are still at the R&D, pilot or demonstration stage (Lange et al., 2018). We deem governance in the MRE sector as involving policy planning, public engagement, and industry development. It bears noting that MRE developments inevitably engenders a complex set of relationships, which then involves multiple stakeholders within and outside coastal communities. Studies undertaken by Lange et al. (2018) and Petrova (2014) relative to wind energy identify pertinent challenges such as fiscal and investor uncertainties, incoherence and unclear regulation and planning at the national level. The planning was more of short-termism instead of the desired long-term horizon. Technological challenges were also identified especially in relation to device development, grid connection and security of energy supply. The planning and regulation failures could be attributed to the absence of nested governance as the core entities were ultra-autonomous to yield to a centralized authority. The net effect is the failed nested governance system which could not provide economic certainty and enhance bottom-up energy transitions. However, the effective cases demonstrated effective combination of multiple stakeholder approach (policy, academic and planning-led), and ultimately resulted in successful industry development and project implementation.

From the New Institutional Economics perspective, institutions are mainly understood as necessary constraints, that is, formal and informal systems of rules—built by human beings to reduce the high degree of uncertainty that characterizes their interactions. North (1990, p. 3) proposes the following definition: "Institutions are the rules of the game in a society or, more formally, are the humanly devised constraints that shape human interactions [structuring] incentives in human exchange, whether political, social or economic". This definition is grossly inadequate the capture the complex intervening production oriented functions of institutions.

The pursuit of successful MRE would require transformative change in structures/organizations, infrastructure, behaviors, governance practices and cultures. Therefore, critical vectors of society's governance landscape would require a critical assessment of the nature of institutions necessary to articulate and achieve the national energy policy priorities. Taking cue from other institutional economists such as Aoki (2001) and Hodgson (2000), institutions have both constitutive and instrumental effects. Besides, these institutions are not pluralistic in orientation from ab initio, rather it takes institutional re-engineering to create such cohesive, interdependence, and indivisibility. In particular, institutions relevant to the development, production, distribution and storage of renewable energy in general are complex and diverse. Additionally, the complexity of that infrastructure defies the capacity and capabilities of a single entity as witnessed by the interplay of manifold institutions advanced economies (Lange et al., 2018). They range from financial, technology, engineering, research, relevant government agencies and civil society. Hence, it is from this multiple stakeholder context that the polycentrism becomes evident as there appears to be no overarching central control. An absence of a potent coordination entity is likely to trigger uncertainty and unpredictability of the outcomes.

The uncertainty issue as argued by Simon (1983) and Stiglitz (2015) stems from the fact that human knowledge and information is incomplete and asymmetrically distributed in the wake of data inaccuracy would adversely affect the quality of decision-making. North (1990) opines that the uncertainty also derives from the "non-ergodic" nature of human domain. On the other hand, the constraining function of institutions is arguably a narrow and less dynamics position to rely on. Hodgson (2006) and Chang and Evans (2005) privilege an enabling and constitutive functions of institutions. Apart from North's constraining market primacy. Chang and Evans (2005) argue that institution's constraints on human behavior rather enable everyone "collectively to do more things" especially the capability to undertake complex forms of coordination within and wider market. The constitutive function of institution lies in Hodgson (2006) postulation of institutions role in shaping behaviors, interests, ideas, values, and motivations of individuals toward a collective goals or to achieve a set national objectives as social structures. Thus, institutions are supposed to provide the requisite incentives for both individual and collective change. Further, these processes of values diffusion are not seamless, as they are bound to engender conflicts, potential setbacks or resistance. Notably, the constitutive role of institutions serves as a software for the instrumental function of empirical articulation of expected practices and policies. The logical implication is that the institutions that undergird renewable energy systems are supposed to undertake both market and non-market functions in coherent, sustainable, and coordinated manner.

Arguably, the extant institutions for Ghana's MRE sector given their disparate nature would likely experience such institutional rigidities. The key factor therefore is the ability to handle such expected and unanticipated shocks. Perusing the NETP depicts a low prioritization of institutional quality apart from few lines that focus on capacity building. There are potential problematic areas germane to institutional responsiveness, coherence, co-ordination and overall dynamism of MRE development based on studies (Stirling, 2015; Lange et al., 2018). The nature and depth of institutional quality has been a significant bane to many energy sectors in the developing economies. This institutional quality has been a bane in notable energy sectors of Ghana. In particular, the country has displayed immense poor outcomes in petroleum refinery as the Tema Oil refinery has been saddled with colossal debts. The Ghana's electricity sector has not been spared with management and operational challenges as well as heavy indebtedness. Logically, the evolving structural and institutional crises in Ghana's energy sector casts legitimate doubts about the Government's cardinal objective of achieving 10% of renewable energy by 2030. This is reinforced by the fact that the 2023 renewable energy production was barely 1% (NETF, 2023).

The relevance of institutional quality cannot be over-emphasized as it is critical to the competitiveness and robustness in managing the transitional processes of the MRE systems and structures. Importantly, quality institutions are necessary to manage short-term shocks which influence the competitiveness of and viability of the private sector in particular (Lebdioui & Bilek, 2021). Currently, the participation of domestic private sector in the Ghana energy sector is limited in terms of competitiveness, scale, and overall quality. Clearly, the private sector would require large investment, high capacity utilization, financing, right mix of incentives, human capital and competitive supplier firms. These structural and institutional factors require a sound policy environment to attract optimal resources.

Strikingly, Ghana's policy environment has not been that promising to the international community relative to development finance in view of the country's overwhelming debt burden. A critical factor which is not adequately considered in the pursuit of energy financing is the role of the nation's macroeconomic management. Two crucial variables of interest that needs to be carefully managed are interest rates and exchange rates. Clearly, in a poor economy with high interest rates clearly serve as a disincentive to the private sector access to project finance. In fact, Ghana's average lending rate of 28% particularly shuts the door to finance to investors and further expose industry to external markets with stringent requirements. This dismal condition invariable makes external financing the inevitable choice for the Government of Ghana relative to project financing. Also, the exchange rate factor from an import-dependent lower middle income country context that Ghana finds itself. The import trap of Ghana has significantly weakened the currency and thereby affecting the cost competitiveness of the renewable energy sector as the case pertains in the overall extractive sector. It bears noting that the MRE sector just like the traditional extractive sector, is likely to heavily rely on imported inputs due to the non-availability of many of the operational components in the domestic markets. Thus, a situation of less cost competitiveness would shore up overall costs and ultimately sub-optimal performance. Also, the NEP (2021) attests that the country produces highly priced electricity.

As adumbrated earlier, the continuous dominance of the thermal plants for electricity generation comes at a very high cost with dire fiscal implications. This dismal situation significantly emanates from the weak exchange rate regime that the Independent Power Producers (IPPs) continue to endure and associated with the dependence on imported petroleum to power the plants shore up their overall cost. Thus, given the indebtedness of Government to the IPPs, the possibility of sending wrong signals to prospective investors in renewable energy sector cannot be ignored. In fact, one can wager that the inability to meet the 10%

renewable energy contribution to electricity by 2020 is significantly attributed to this macroeconomic malaise and its correlating downside effects. Indeed, this macroeconomic maelstrom adds to the uncertainty of the polycentric governance of renewable energy in Ghana as the State's control of the sector cannot be fully guaranteed against the other players. The overarching cautionary tale is that in reifying institutions one must guard against linear thinking, uniformity assumptions, non-static/evolving nature of institutions, as well as the political nature of institutions. Also, context matters in the form and content of any institution and institutional performance, and that their articulation is imbued with uncertainties.

Institutions for the Marine Renewable Energy Development

As discussed above the relevance of institutions for the renewable energy sector in general is firmly established. However, the critical area of importance relates to what or which type of institutions are apposite to the development, viability and sustainability of the renewable sector. Taking cue from Abramovitz (1995, p. 35) social capability factor as comprising two complementary elements: first "people's basic social attitudes and political institutions" and "ability to exploit modern technology. In essence, the quest for structural transformation in any economy via industrialization and other sectoral change would need formation of specific institution. This orientation resonates with a perspective of development as "a process of production transformation led by the process of collective capabilities and resulting in the creation of good quality jobs and sustainable structural change (Andreoni & Chang, 2017, 173). Admittedly, there are other vital institutions undergirding the structural transformation for any economy. Indeed, institutions for maintenance of law and order to ensure social stability and cohesion are necessary, however, discussing these wide-ranging institutions is clearly beyond the remit of this chapter. Hence, we will discuss some of the vital economic and industrial related institutions that are preconditions or necessary for an effective MRE. The focus in this case is to approach institutions in a disaggregated manner by focusing on specific institutions needed to produce the social capabilities to accelerate the energy sector transformation. The types of institutions needed for the renewable energy include the following: productive capabilities, production, corporate governance, industrial financing, industrial change and restructuring. However, for the sake of brevity, discussion focuses on three institutions namely: macroeconomic, production and production capabilities.

Production Institutions

Hitherto, the production system from the industrial revolution era started with the factory system as a source of wealth creation as seen by Smith and critiqued by Marx. However, the factory system has evolved exponentially from the general, specialized/skilled via Taylorism and internationalized within the globalization context. Additionally, the current production chains systems are highly technologically driven, which is a challenge for technologically less endowed developing economies to compete. It is vital to note that these different epochs of production system were shaped by specific institutions. The Fordist era mass production has its specific institutional framework as much same with the current global value chains. Indeed, specific institutions were responsible for the creation of the different production systems in the East Asian economies to achieve those unprecedented long and high growth rates. In the same vein, developing countries such as Ghana must develop the production systems for its vital sectors, in this case the renewable energy by adapting to established strategies.

Additionally, the current production system largely driven through the global value chain (GVC) has provided developing countries opportunities to participate in essential components of the chain. Therefore, taking cognizance of the increasing influence of GVCs, in particular the MRE sector, domestic producers in developing economies must explore ingenious ways to learn to climb up the value chain through the enhancement of their local productions systems. Clearly, this would require the domestic producers or potential investors to invest significant resources to mastering technologies, managerial techniques, and worker productive skills.

Macroeconomic Institutions

The relevance of the nature of the macroeconomy to the MRE sector is unquestionable. As discussed earlier the macroeconomy question sparks a rethinking of the economic management policy strategy of the government. This is due to the fact that notwithstanding the many years of economic reforms, the macroeconomy of star reforming countries such as Ghana is yet to stimulate the renewable energy sector. Clearly, the over-reliance of a liberal macroeconomic policy regime can be counter-productive in both the short and long run as seen in Ghana's case with industrialization. Importantly, there is a need to reshape the central bank away from its mere independent status into real functionality to provide: progressive economic stability in much broader sense; ensure that interest rates are low; ensure that the currency is not over-valued; and targeted and subsidized funding of project of national interest (Epstein, 2007).

Given the low tax base of the Ghana's economy it would be imperative to establish budgetary institutions to undertake a broad-based tax system engineered with minimum prospect of tax evasion as a way of shoring up domestic resource mobilization. Additionally, there is a need to institute budgetary rules that promotes active Keynesian fiscal policy to ostensibly support government expenditure on energy (Andreoni and Chang, 2017). As a matter of fact, the minuscule budgetary expenditure as against the budget deficit or public debt imprimatur has not been helpful in closing the energy gap.

Institutions for Capabilities Building

Clearly, the evolution in the production system as discussed above attests to the deliberate processes that ensured their manifestations. These processes of leapfrogging the industrial chain were as a result of deepening of technological diffusion. It necessary to note that change in industrial output or productivity or diversification emanate inter-play of firm's internal restructuring or strategies by investing more in research and development to build their capabilities. Indeed, firms are deemed as learning entities as learning undergirds its innovation prospects and overall competitiveness. Significantly, the MRE is a knowledge intensive sector which inevitable requires an effective nexus between industry-academic-policy. Qiu et al. (2019) and Lange et al. (2018) show the enormous degree of collaboration between the marine industry players and the various academic and research institutions worldwide. Their studies further points to varying degrees of knowledge sharing among the firms and policy makers in the United States, United Kingdom, Denmark, Norway, and China. The various countries have established diverse research institutions with incentives for knowledge diffusion as a way of optimizing the benefits of the MRE.

Concluding Remarks

The quest for Ghana's energy transition objective in line with national and global commitments is indeed a daunting task as against the current development outlook of the country. Besides, the chapter invites scholarship into the critical relational processes vital to MRE and energy transition in general. From our perspective, the disproportional research focus on technical and scientific components of the transition and minuscule attention accorded to institutional and governance elements tends to reduce the complexity nature of the energy transition. Urgently, needed is transdisciplinary research to understand the complexities. Notably, the polycentric governance processes also unearths this unavoidable complexity in the articulation of renewable energy efforts. Additionally, the presence of poly-crisis (debt crisis, energy deficit, conflict, food deficit etc.) in the global system in a significant way is likely to increase the vulnerabilities of developing countries effort at energy transition. This further nudges for a critical assessment of the international system, particularly how bilateral and multilateral arrangements, determine developing countries prospects of optimizing MRE.

References

Abramovitz, M. (1995). The elements of social capability. In B. H. Koo, & D. H. Perkins (Eds.), *Social capability and long-term economic growth*, Trans. (pp. 19–47). Macmillan.

Agyekum, E. B. (2020). Energy poverty in energy rich Ghana: A SWOT analytical approach for the development of Ghana's renewable energy. *Sustainable Energy Technology, 40*.

Andreoni, A., & Chang, H.-J. (2017). Bringing production and employment back into development. *Cambridge Journal of Regions, Economy and Society, 10*, 173–187.

Aoki, M. (2001). *Toward a comparative institutional analysis*. MIT Press.

BP Statistical Review of World Energy. (2022). Retrieved from http://www.bp.com/statisticalreview.

Chang, H.-J., & Evans, P. (2005). The role of institutions in economic change. In G. Dymski & S. Da Paula (Eds.), *Reimagining growth*. Zed Press.

Cui, Y., & Zhao, H. (2023). Marine renewable energy project: The environmental implication and sustainable technology. *Ocean and Coastal Management, 232*.

Darko, D., Kpessa-Whyte, M., Obuobie, M., Siakwah, P., Torto, O., & Tsikata, D. (2019). The context and politics of decision-making on large dams in Ghana: An overview. University of Manchester, Future Dams Project, Working Paper 001.

Emission Database for Global Atmospheric Research (EDGAR). (2022).

Epstein, J. (2007). Central banks as agents of economic development. In H-J. Chang (Ed.), *Institutional change and economic development*. United Nations University Press, and Anthem Press.

Garmestani, A. S., & Benson, M. H. (2013). A framework for resilience-based Governance of Social-Ecological Systems. *Ecology and Society, 18*(1), 9.

Ghana Statistics Service. (2021). *Population and Housing Census*.

Government of Ghana, "Renewable Energy Act 2011: Act 832", 1–27.

Gupta, A., & Ferguson, J. (1999). Discipline and practice: The field as site, method, and location in Anthropology. In A. Gupta & J. Ferguson (Eds.), *Anthropological locations: Boundaries and grounds of a field science* (pp. 1–46). University of California Press.

Hall, A., & Branford, S. (2016). Development, dams and dilma: The Saga of Belo Monte. *Critical Sociology, 38*(6), 851–862.

Hickey, S. (2013). *Thinking about the politics of inclusive development: Towards a relational approach*. ESIS Working Paper 1. University of Manchester.

Hodgson, G. (2000). *Structures and institutions: Reflections on institutionalism, structuration theory and critical realism*, mimeo, The Business School, University of Hertfordshire.

Hodgson, G. (2006). What are institutions? *Journal of Economic Issues, 40,* 1–25.

Lange, M., Page, G., & Cummins, V. (2018). Governance challenges of marine renewable energy developments in the U.S. creating the enabling conditions for successful project development. *Marine Policy, 90,* 37–46.

Lavers, T., Gebresenbet, F., and Terefe, B. (2024). Political Vulnerability and the Origins of the EPRDF's Dams Boom, in Lavers, T. (ed) *Dams, Power and the Politicsm of Ethiopia's Renaissance*, pp.60-87. Oxford: Oxford University Press

Lebdioui, A., & Bilek, P. (2021). *Do forward linkages reduce or worsen dependency in the extractive sector?* Background paper.

Leftwich, A. (2005). Development studies and the rediscovery of social science. *New Political Economy, 10,* 573–607.

Mannaseh, R., Sannasiraj, S. A., McInnes, K. L., Sundar, V., & Jalihal, P. (2017). Integration of wave energy and other marine renewable energy sources with the needs of coastal societies. *The International Journal of Ocean and Climate Systems, 8*(1), 19–36.

National Energy Policy. (2021).

National Energy Transition Framework. (2023).

NEP (2022) *Ghana National Energy Policy*, Ministry of Energy

North, D. (1990). *Institutions*. Cambridge University Press, Cambridge.

Okereke, C. (2024). Degrowth, green growth, and climate justice for Africa. *Review of International Studies, 50*(5), 910–920.

Ostrom, V., Tiebout, C. M., & Warren, R. (1961). The organization of government in metropolitan areas: A theoretical inquiry. *American Political Science Review, 55*(4), 831–842.

Parliament of the Republic of Ghana. (2011). Act 832.

Petrova, M. A. (2014). Sustainable communities and Wind Energy Project Acceptance in Massachusetts. *Minnesota Journal of Law, Science and Technology, 15,* 529–553.

Qiu, S., Liu, K., Wang, D., & Liang, F. (2019). A comprehensive review of ocean wave energy research and development in China. *Renewable and Sustainable Energy Reviews, 113.*

Simon, H. (1983). *Reason in human affairs*. Stanford University Press.

Stephan, M., Marshall, G., & McGinnis, M. (2019). An introduction to polycentricity and governance. In A. Thiel, W. Blomquist, & D. Garrick (Eds.), *Governing complexity.* Cambridge University Press.

Stirling, A. (2015). *Developing 'Nexus Capabilities": Towards transdisciplinary methodologies.* University of Sussex, Discussion paper, June 29–30.

Stiiglitz, J.E. (2015). Industrial Policy, *Learning, and Development WIDER Working Paper, 149*, UNU-WIDER and the Korean International Cooperation Agency, Helsinki, Finland.

World Bank. (2023). *Reality check: Lessons from 25 policies advancing a low-carbon future.* Climate change and Development Series. World Bank.

World Bank. (2024). *Within reach: Navigating the political economy of decarbonization.* Climate change and Development Series. World Bank.

8

Renewable Energy Innovations in Urban Sustainability: Water and Energy Synergies

Anna Minakova

Abstract This chapter investigates the integration of renewable energy solutions in urban infrastructure, emphasizing their potential to revolutionize water and energy management systems and water/energy management indicators in building standards such as BREEAM, LEED. The intersection of sustainable urban development and renewable energy adoption presents a roadmap toward creating more resilient and efficient urban environments.

Keywords Blue zones · Renewable energy · Water consumption · Sustainable urbanism · Sustainable building · Water management

Introduction

Rapid urbanization is reshaping cities into complex systems with soaring demands for water and energy. Balancing sustainability, resilience, and livability requires integrating renewable energy, advanced water management, and digital innovations. It puts cities at the center of sustainability challenges and solutions. Water and energy play a fundamental role in urban life. The interdependence between water and energy systems is

© The Author(s), under exclusive license to Springer Nature
Switzerland AG 2025
Y. Ermolaeva et al., *Rethinking Water and Energy for a Sustainable Future*, Palgrave
Studies in Climate Resilient Societies, https://doi.org/10.1007/978-3-032-04485-3_8

a crucial consideration for sustainable urban development. Energy is needed to extract, treat, distribute, and heat water. Water is essential for generating energy, especially for cooling thermal power plants and hydropower production. Inefficiencies or disruptions in one system can affect the other, increasing risks to urban resilience.

International frameworks such as LEED (Leadership in Energy and Environmental Design (LEED) is a green building certification program used worldwide. Developed by the non-profit U.S. Green Building Council (USGBC), it includes a set of rating systems for the design, construction, operation, and maintenance of green buildings, homes, and neighborhoods. The program aims to help building owners and operators become more environmentally responsible and use resources more efficiently) Cities and BREEAM (The Building Research Establishment Environmental Assessment Method, or BREEAM, was first published by the Building Research Establishment in 1990 and is widely recognized as the world's most established method for assessing the sustainability of buildings) Communities offer structured pathways for incorporating sustainability into urban planning and operations. These standards establish benchmarks for resource efficiency, renewable energy integration, green infrastructure, and stakeholder involvement. However, the effectiveness of these frameworks depends on local adaptability, the ability to scale solutions in a fair and equitable manner, and meaningful engagement with the community.

This chapter explores the incorporation of renewable energy sources into urban infrastructure, with a specific focus on their ability to revolutionize water and energy management systems. Through an examination of the intersection between sustainable urban development, technological advancements, and international best practices, a roadmap for building more resilient, efficient, and livable cities is presented. Special attention is paid to the often overlooked potential of hydro and marine energy technologies, as well as the systemic challenges cities face when managing water consumption and conservation.

Accelerating Urbanization and Resource Management

The twenty-first century has seen an unprecedented wave of urbanization, transforming cities into densely interconnected networks of human activity, infrastructure, and resource flows. Currently, more than 55% of the world's population lives in urban areas, and this figure is expected to rise to nearly 70% by 2050 (United Nations, 2018). This demographic shift is a fundamental transformation in how humanity interacts with finite resources on our planet. Cities, which occupy less than 3% of Earth's land (Martin, 2023), are responsible for more than 70% (World Bank - Cities Key to Solving Climate Crisis, 2023) of global carbon emissions and consume over two-thirds of energy (IEA, 2023).

Urban water management faces significant challenges. Aging infrastructure in developed countries leads to water losses of up to 30% (Kingdom et al., 2006). Climate change is exacerbating these issues by increasing the frequency and intensity of droughts, floods, and heatwaves. These events disrupt both water supply and energy systems (IPCC, 2023). Additionally, urban sprawl is increasing impervious surfaces, reducing groundwater recharge, and increasing stormwater runoff. This elevates flood risk and contributes to water pollution (Team, 2025).

A further complication arises from the lack of alignment between the interests of resource companies and those of consumers. Utilities may have financial incentives to maintain or increase extraction rates, as their revenues are often linked to the amount of water or energy they sell. At the same time, residents are encouraged to adopt water and energy saving behaviors, but without systematic support or visible outcomes, their efforts can seem futile. This discrepancy erodes public confidence and compromises the effectiveness of supply-side management approaches.

One of the major challenges to sustainable water management and energy planning in cities is the lack of coordination between different agencies responsible for these areas. Urban resource systems are managed by a variety of organizations, including water utilities, energy companies, environmental agencies, and planning bodies, each with their own objectives, budgets, and priorities. This lack of integration often leads to inefficient use of resources and missed opportunities for comprehensive solutions.

Transparency is essential for building public trust and promoting progress in urban sustainability. However, many utilities and resource companies fail to publish clear and time-bound plans for reducing water extraction and energy consumption. They also do not regularly disclose their progress, which frustrates residents who are encouraged to conserve resources but see little evidence of change. This lack of transparency makes it difficult for regulators and policymakers to identify best practices and hold underperforming utilities accountable.

Urbanization often exacerbates inequalities in access to water and energy, particularly in developing countries. In cities, many people live in informal settlements without piped water or grid electricity, resorting to expensive or unsafe alternatives. In many sub-Saharan African cities, residents in low-income and informal settlements often pay substantially higher prices for water than those connected to municipal systems, primarily due to reliance on informal vendors and communal supplies. Studies in cities such as Dar es Salaam, Blantyre, and Harare show that water costs can consume between 11% and over 100% of household income for some low-income families, far exceeding the recommended affordability threshold of 5% (Mitlin, 2019). This disparity highlights the urgent need for policies that improve access to affordable, safe water and address the financial barriers faced by the urban poor. Even in high-income countries, affordability of essential services like water is a growing concern. Detroit conducted residential water shut-offs due to non-payment, with multiple news reports documenting that these actions disproportionately affected low-income residents (Widespread Water Shut-Offs in US City of Detroit Prompt Outcry from UN Rights Experts, 2014). The cumulative impact of water shut-offs in Detroit has drawn national attention to issues of water affordability and equity.

Access to new technologies like rooftop solar panels and smart meters could help reduce these inequalities. Water-saving devices are also unevenly adopted, with affluent neighborhoods leading the way, leaving marginalized communities behind in terms of innovation. This "digital divide" can exacerbate social inequality unless addressed through targeted subsidies, inclusive planning, and community engagement.

Leveraging digitalization, transparency, stakeholder engagement is essential to optimize resource use. Technological innovation is transforming the way cities manage water and energy resources. With the use of smart metering and internet of things (IoT) sensors, utilities can detect leaks and monitor consumption in real-time (UNESCO, 2020). This helps to reduce water loss and energy waste, as well as optimize pumping schedules and energy dispatch with the help of advanced analytics and artificial intelligence (AI).

While these advancements provide powerful tools for increasing efficiency and system responsiveness, they cannot eliminate the broader challenges faced by urban infrastructure. Even the most advanced management systems must deal with external pressures that can threaten the reliability and safety of water and energy supplies. Climate change and vulnerabilities in infrastructure make matters more complicated, as climate-related risks such as extreme weather events and rising sea levels can disrupt water supplies, while aging infrastructure can lead to breakdowns.

To address these issues, it is essential to implement policies that prioritize equity, sustainability, and resilience in water management. These policies should aim to ensure that all communities have access to safe and reliable water supplies, regardless of their location or socioeconomic status. Additionally, they should focus on reducing climate change impacts and promoting the long-term sustainability of water resources. By implementing such policies, we can create a more resilient and equitable water system that can withstand the challenges of the future.

However, the urgency of these policy measures is emphasized by the growing evidence of how climate change and ageing infrastructure are already affecting cities around the world. For example, coastal aquifers in cities such as Miami and Shanghai are facing an increasing problem with saltwater intrusion. The 2019 water crisis in Chennai is a stark reminder of how urbanization and over-extraction of groundwater, combined with unpredictable monsoons, can leave millions of people without reliable water supply for weeks ("How Chennai, One of the World's Wettest Major Cities, Ran out of Water," 2021). In California the drought of 2022 led to utility companies reducing hydroelectric power production and implementing rolling power outages as reservoirs ran dry (Short-Term Energy Outlook Supplement: Drought Effects on California Electricity Generation and Western Power Markets, 2022). Urban infrastructure, often built decades ago, often lacks the necessary equipment to deal with extreme weather conditions. In London, for instance, record rainfall in 2021 caused flooding in underground train stations and residential areas, illustrating the limitations of the current stormwater system (London Floods Highlight Failure to Address Climate Change & Its Consequences, 2021). This transition clearly connects the need for strong policies to the real-life consequences of climate change and infrastructure risks, creating a logical and coherent flow.

One of the most promising innovations in energy management in the city is the use of distributed energy storage in multi-unit buildings. Urban areas face significant peak loads, particularly in residential areas, and batteries installed within buildings can help to store energy during off-peak periods and discharge it during peak times, reducing strain on the grid and delaying costly infrastructure upgrades.

This not only benefits the grid but also residents, who can participate in demand response programs without compromising their comfort. In Germany, regulatory reforms have allowed home storage systems to feed excess energy back into the grid, increasing flexibility (Blathnaid O'Dea, 2024). Similarly, community solar and storage projects in California are projected to save billions and reduce emissions (Markham, 2025).

However, there are still challenges to overcome. What incentives are best to encourage adoption? How can we ensure equity so that all residents benefit, rather than just early adopters or wealthy households? All these questions are waiting for us soon, and we need to find answers to them now.

The interconnectedness of water and energy systems in cities presents both challenges and opportunities. Water is essential for every stage of the energy cycle, from extraction and treatment to distribution and wastewater management. Energy, in turn, plays a crucial role in water production, particularly for cooling thermal power plants and hydropower generation. Disruptions in one system quickly affect others.

Despite the connections between water and energy systems, they are often managed separately, missing opportunities to improve efficiency. Integrated planning, like co-locating infrastructure or using wastewater for heating, is the exception rather than the rule.

How LEED Cities and BREEAM Communities Address Urban Water and Energy Challenges

LEED Cities (2024) and BREEAM Communities (BREEAM Communities—BREEAM—Liferay DXP, n.d.) are internationally recognized frameworks that guide cities and communities toward sustainability, resilience, and resource efficiency. LEED Cities focuses on holistic urban sustainability, including energy and water efficiency, climate resilience, equity, and transparency. BREEAM Communities is a science-based standard for masterplanned developments, emphasizing integrated planning, green infrastructure, and community engagement. Both standards require cities and developments to set measurable targets and involve stakeholders to address the most pressing urban water and energy challenges.

LEED Cities and BREEAM Communities require cities and developers to assess and reduce water losses from aging infrastructure, through measures such as leak detection, smart meters, and regular maintenance. They also address climate change risks by requiring resilience planning, including strategies for droughts, floods, and heatwaves, as well as green infrastructure and stormwater management.

First, both standards require cities and developers to assess and reduce water losses from aging infrastructure through measures such as leak detection, smart meters, and regular maintenance.

Second, they address climate change risks by requiring resilience planning—cities must demonstrate strategies for preparing for droughts, floods, and heatwaves, including the use of green infrastructure and stormwater management.

Third, these frameworks promote integrated resource management by breaking down barriers between water, energy, and urban planning sectors. This is done through requirements for cross-sector planning, stakeholder involvement, and transparent reporting.

Fourth, LEED Cities and BREEAM Communities establish criteria for equitable access to ensure that all communities, including informal settlements and low-income areas, benefit from improved water and energy services and digital solutions.

Fifth, transparency and accountability are built into both standards: cities are required to publish clear, time-bound targets and report on progress, helping to rebuild public confidence and enable regulatory oversight.

Sixth, the standards promote the adoption of innovative technologies such as IoT sensors, smart meters, distributed energy storage, and water reuse systems, while also addressing the digital divide through inclusive planning and targeted subsidies.

Finally, by mandating integrated water-energy planning and resilience measures, these standards assist cities in adapting their infrastructure to extreme weather events and future uncertainties, ensuring long-term sustainability for all residents.

Urban challenges and how LEED Cities and BREEAM Communities address them

Urban challenge	LEED Cities Criterion	BREEAM Communities Criterion
Aging infrastructure and water loss	WE Prerequisite: Integrated Water Management: - Water Availability Assessment - Water Demand - Water Supply	RE 03—Water strategy: Infrastructure monitoring, water loss reduction
Climate change impacts (droughts, floods, heat)	WE Credit: Stormwater Management: Reduction of runoff volume, prevent erosion, flooding and recharge groundwater EN Prerequisite: Energy and Greenhouse Gas Emissions Management: Moving towards a zero emissions city and reduce environmental and economic harms associated with excessive energy use EN Credit: Renewable Energy: To reduce the environmental and economic harms associated with fossil fuel energy and reduce Greenhouse Gas emissions by increasing self-supply of renewable energy, use of grid-source renewable energy technologies and carbon mitigation projects EN Credit: Low Carbon Economy: To progress towards a low carbon economy by decoupling economic growth of the city from greenhouse gas emissions	SE 03—Flood risk assessment: A site-specific flood risk assessment is carried out in accordance with current best practice and planning policy RE 03—Water strategy: The developer engages with water suppliers, the local authority and the appropriate regulatory body to develop overall water consumption targets for the development taking account of: - The future predicted availability taking climate change into account - The predicted water demand for the area resulting from growth and climate change SE 13—Flood risk management: The recommendations of the appropriate statutory bodies within the site-specific flood risk assessment are incorporated into the masterplan

(continued)

(continued)

Urban challenge	LEED Cities Criterion	BREEAM Communities Criterion
Urban sprawl and stormwater management	WE Credit: Stormwater Management: Reduction of runoff volume, prevent erosion, flooding and recharge groundwater WE Credit: Wastewater Management: Use treated wastewater to meet the city water demand	LE 03—Water pollution: A comprehensive and up-to-date drainage plan of the site is produced by a appropriately qualified professional The drainage plan will be made available to the authority responsible for maintaining the drainage infrastructure and future development occupiers Measures are put in place to avoid any potential water pollution LE 06—Rainwater harvesting: Rainwater collection systems are designed and specified
Misaligned incentives/ utilities vs. consumers	IP Prerequisite: Integrative Planning and Design Process: Design Charrette: As early as practical and preferably before master plan approval, conduct a design charrette with the project team as defined above and representatives of the citizens who will get impacted by the project	GO 01—Consultation plan: Members of the local community and appropriate stakeholders have been identified for consultation

(continued)

(continued)

Urban challenge	LEED Cities Criterion	BREEAM Communities Criterion
Fragmented governance	INTEGRATIVE PROCESS (IP): Assemble and convene an interdisciplinary, cross-departmental project team. Include diverse team members from at a minimum three of the following areas of expertise - Development Authority - Urban/Master Planning and Design - Engineering (Energy and Power; Hydrology; Transportation; Waste) - Economic Development - Urban Ecologist, Biologist or Landscape Architect - Construction Management - Human Services - Education/School Board - Sustainability/Resilience Officer - Data Officer/Information Technology	SE 10—Adapting to climate change: Evidence has been used from the local authority and statutory bodies to understand the known and predicted impacts of climate change on the site GO 01—Consultation plan: Members of the local community and appropriate stakeholders have been identified for consultation SE 09—Utilities: The following service providers have committed to coordinate the installation of related infrastructure, as relevant: - Gas - Electricity - Water/sewerage - Telecommunications/internet - Heat and cooling (where relevant)

(continued)

(continued)

Urban challenge	LEED Cities Criterion	BREEAM Communities Criterion
Lack of transparency		GO 01—Consultation plan: Members of the local community and appropriate stakeholders have been identified for consultation GO 02—Consultation and engagement: Good practice consultation methods are used to engage members of the community and appropriate stakeholders in the process of designing development proposals GO 04—Community management of facilities: To support communities in active involvement in developing, managing and/or owning selected facilities
Inequitable access to water/energy	EN Prerequisite: Power Access, Reliability and Resiliency: Provision of safe, secured, reliable, resilient, and equitable access to power	RE 03—Water strategy: The developer engages with water suppliers, the local authority and the appropriate regulatory body to develop overall water consumption targets for the development taking account of: - The current availability of water and demands in the area RE 01—Energy strategy: To recognise and encourage developments designed to minimise operational energy demand, consumption and carbon dioxide emissions

(continued)

(continued)

Urban challenge	LEED Cities Criterion	BREEAM Communities Criterion
Digital divide in technology adoption	WE Credit: Smart Water Systems: Smart Water Metering IN Credit: Innovation: To encourage cities to achieve exceptional or innovative performance	SE 17—Training and skills: The developer consults with the community, local businesses, training providers and relevant authorities to identify training and skills initiatives that would be beneficial to the local area Inn 01—Innovation: Any new technology, design, planning or construction method or process can potentially be recognised as 'innovative'
Infrastructure vulnerability	QL Credit: Affordable Housing: To provide access to housing at reasonable costs to sections of the society which need assistance QL Credit: Emergency Management and Response: To create sufficient capacity and capability to respond to emergency incidents and reduce its impact on human life/health	SE 15—Inclusive design: Emergency egress strategies SE 03—Flood risk assessment: An emergency plan is established in the event of flooding SE 10—Adapting to climate change: Demonstrate how the risks will be managed and reduced through the use of 'win–win' measure

(continued)

(continued)

Urban challenge	LEED Cities Criterion	BREEAM Communities Criterion
Disconnected water-energy management	EN Prerequisite: Energy and Greenhouse Gas Emissions Management: Moving towards a zero emissions city and reduce environmental and economic harms associated with excessive energy use WE Credit: Smart Water Systems: Smart water systems can link together multiple systems within a network to share data across platforms. Considering many of the common challenges faced by utilities, including leak management, regulation compliance, and customer service, utilities can improve performance by integrating systems in a manner that tracks and highlights specific problem areas	RE 01—Energy strategy: To recognise and encourage developments designed to minimise operational energy demand, consumption, and carbon dioxide emissions RE 03—Water strategy: A water strategy is prepared to manage water demand on the development site to meet the above consumption targets. The strategy includes: - Actions to minimise the predicted use on the development; and maintain this in the future - Ownership and maintenance of any shared facilities - Design options to reduce the water demand in landscaping, any other predicted water use and on-site collection/storage opportunities - Targets for water use in residential and non-domestic buildings in the development site

Cited and Adapted from LEED Cities (2024) and BREEAM Communities—BREEAM—Liferay DXP (n.d.)

The city of Greensboro, North Carolina, has made significant progress in sustainable urban planning and management. This is evident in its LEED certification as a LEED City (USGBC, 2020). In terms of water management, the city has achieved a water efficiency score of 44 out of 100. The estimated daily per capita water consumption is 85.60 gallons in 2018.

Greensboro ensures equitable access to water and sanitation for all residents. Its public water distribution system and central sewage system cover over 99% of the city's buildings. Only a small percentage of buildings (less than 1%) rely on private systems regulated by the local health department.

Both drinking water and wastewater are consistently maintained to meet US EPA standards. This ensures the safety and quality of water supply and treatment for residents.

Greensboro uses an integrated water management approach that assesses total available water resources, sectoral demand, and supply from various sources, including reclaimed water. In 2019, the city's Water Resources Department conducted a comprehensive water loss audit, leading to the development of a leakage management plan to improve water distribution efficiency.

The city has also prioritized energy efficiency in its energy management. Greensboro has an impressive energy performance, with a per capita greenhouse gas emission of 11.86 tons of CO_2 (2016). To reduce energy use, the city has implemented stringent efficiency standards for streetlights, with 80% of Duke Energy-leased fixtures meeting these standards. Additionally, the city is replacing older fixtures with LED lights, with over 4,000 upgrades already completed. Floodlights are not permitted in public areas by city statute, further reducing energy use.

Grid harmonization is another important achievement. Duke Energy manages Greensboro's electricity infrastructure and provides dynamic pricing programs for residential, commercial, and industrial customers, promoting load shifting to reduce stress on the grid. These integrated efforts in water and energy show Greensboro's commitment to sustainability, resilience, and equal access for all residents through the LEED Cities program.

Addressing urbanization pressures requires sustainability at the community scale. BREEAM Communities offers a science-based certification for masterplanned developments, embedding renewable energy, water management, and green infrastructure into urban design.

London's Battersea Power Station redevelopment illustrates this well. Certified under BREEAM Communities, it integrates renewable energy systems and sustainable water management to create a low-carbon, resilient neighborhood (Chantry & Turcu, 2024). But what are the economic trade-offs? How do upfront costs compare to long-term savings? And how do these communities navigate social equity to ensure benefits reach all residents?

This large-scale mixed-use redevelopment project, certified according to BREEAM Communities, aims to create a sustainable and low-carbon neighborhood. The project has created over 20,000 permanent jobs since its inception, including more than 3,000 direct and indirect construction positions. In addition, 1,074 residents have been employed since 2013 (GENERATING OPPORTUNITIES BATTERSEA POWER STATION LOCAL EMPLOYMENT AGREEMENT ANNUAL REPORT FY, 2022/23, n.d.).

The development has invested heavily in the local supply chain, spending nearly £4.9 million within the local boroughs. Of this amount, £1.9 million has been directed toward small and medium-sized businesses, boosting the local economy. It is estimated that when complete, this development will increase annual spending in Wandsworth by £42.9 million.

While upfront costs are high due to restoration and infrastructure investment, long-term economic benefits include job creation, increased local spending, and enhanced property values. The project also supports social integration and community development through the Battersea Power Station Foundation, which has committed millions in grants to local groups.

However, there is a challenge in balancing the high initial capital investment with affordable housing and ensuring that existing communities are not displaced. This is an important consideration for any large regeneration project.

Hydro and Marine Energy: Untapped Urban Potential

Transitioning from the successful examples of LEED Cities certification in Greensboro and the BREEAM Communities approach in London's Battersea Power Station, it becomes clear that urban sustainability frameworks are evolving to address a wide range of energy and water challenges. While the potential of hydro and marine energy in urban areas is often underappreciated, it is becoming increasingly important to consider these sources as cities strive to diversify their energy portfolios and enhance resilience. By taking a closer look at hydro and marine power, cities can unlock new opportunities for sustainable energy production and contribute to a more resilient and sustainable future.

While solar and wind energy have received most of the attention in discussions about urban renewable energy, hydro and marine technologies also offer several benefits that are often overlooked. For example, mini-hydro power plants can be installed in existing urban waterways, providing local, reliable, and clean electricity with minimal environmental impact. The restoration of a historic dam on the Turin's Regio Parco Canal is an excellent example. By converting the unused infrastructure into a mini-hydro plant, the city produces 1.6 GWh of electricity each year, enough to power nearly 600 homes, while improving flood management and boosting local biodiversity (Comino et al., 2020).

The benefits of mini-hydropower projects extend far beyond energy production. These installations can help revitalize urban infrastructure, promote ecological restoration, and decrease dependence on distant energy sources. Additionally, they offer a degree of reliability and stability that some renewable energy sources lack, making them an important addition to solar and wind energy in a balanced urban energy mix.

Practical examples further illustrate the potential of urban hydro and marine power. In New York City, the Roosevelt Island Tidal Energy project harnesses the power of tides in the East River to provide clean energy to the city's grid (Roosevelt Island Tidal Energy Project, New York, n.d.). This project demonstrates that with the right support from government and community involvement, hydro and marine energy can successfully operate in densely populated areas.

Despite these advantages, there are several challenges that need to be addressed. Regulatory barriers, complex permitting processes, and concerns about the impact on aquatic ecosystems can all slow down project development. Additionally, social acceptance is crucial, as local communities must be actively involved in the planning process and share benefits to ensure fair outcomes. Furthermore, integrating small-scale hydro and marine energy with existing urban infrastructure requires careful technical and financial planning to ensure a successful implementation.

Nevertheless, the potential of hydro and marine energy is too significant to ignore. As cities continue to innovate and seek new ways to become more sustainable, it is essential to pay more attention to and invest in these sources of energy. By drawing on the lessons learned from successful LEED and BREEAM projects, where integrated planning, stakeholder engagement, and transparent reporting have been proven effective, urban leaders can unlock the potential of water-based renewable energy. This approach helps diversify the energy mix and strengthens urban resilience and supports ecological restoration. It also brings tangible benefits to local communities.

To summarize, expanding the focus on hydro and marine energy will help cities achieve a more resilient, reliable, and sustainable future. As urbanization continues to accelerate, these underutilized resources have the potential to become key components of low-carbon, resilient cities.

Conclusion: Toward Integrated Urban Sustainability

Building on the recognition of the importance of the synergy between renewable energy, water management, and digital innovation for sustainable urban futures, we understand that a comprehensive approach is needed to address the challenges that cities face today. Distributed energy storage solutions, such as battery systems in multi-unit buildings, help stabilize urban energy grids and empower residents to participate in demand response programs. This increases both resilience and community engagement.

Community-based sustainability certifications like LEED Cities and BREEAM Communities provide structured pathways for integrating resource efficiency, stakeholder involvement, and transparency into urban planning. They set measurable targets for water and energy use, climate resilience, and equity. Innovative technologies, such as mini-hydropower plants and IoT sensors, hold great potential for optimizing resource usage and reducing losses from aging infrastructure. However, the implementation of these solutions presents important challenges related to economic feasibility, regulatory compliance, and social equity.

Without targeted subsidies and inclusive policies, as well as robust community engagement, there is a risk that these technological advancements could exacerbate existing inequalities and leave marginalized groups behind in the transition to more sustainable cities.

Learning from successful case studies, such as Greensboro's LEED City certification, can help urban planners and policymakers better understand how to apply international best practices in their local contexts. These case studies demonstrate tangible progress in areas such as water efficiency, equitable service coverage, and safety standards compliance.

Ultimately, the path forward requires a commitment to implementing innovations in ways that are both fair and efficient. This includes prioritizing policies to ensure access to safe water and clean energy for all residents, investing in resilient infrastructure, and fostering meaningful participation from stakeholders. By learning from leading cities and utilizing the full potential of digital and renewable technologies, we can create urban environments that are not only sustainable and livable, but also fair and inclusive for future generations.

References

Blathnaid O'Dea. (2024, December 4). *Germany to lift restrictions on home storage systems discharging into the electricity grid—Energy Storage.* Energy Storage. Retrieved from https://www.ess-news.com/2024/12/04/home-storage/.

BREEAM Communities—BREEAM—Liferay DXP. (n.d.). *BREEAM*. Retrieved from https://breeam.com/standards/communities.

Chantry, W., & Turcu, C. (2024). Sustainability power to the people: BREEAM Communities certification and public participation in England. *Discover Sustainability, 5*(1). https://doi.org/10.1007/s43621-024-00473-2

Comino, E., Dominici, L., Ambrogio, F., & Rosso, M. (2020). Mini-hydro power plant for the improvement of urban water-energy nexus toward sustainability—A case study. *Journal of Cleaner Production, 249*, Article 119416. https://doi.org/10.1016/j.jclepro.2019.119416

EA. (2023). *World energy outlook 2023—analysis.* IEA. Retrieved from https://www.iea.org/reports/world-energy-outlook-2023.

GENERATING OPPORTUNITIES BATTERSEA POWER STATION LOCAL EMPLOYMENT AGREEMENT ANNUAL REPORT FY 2022/23. (n.d.). Retrieved from https://batterseapowerstation.co.uk/content/uploads/2023/07/Generating-Opportunities-2023.pdf.

How Chennai, One of the world's Wettest Major cities, Ran out of Water. (2021, February 4). *The Economic Times*. Retrieved from https://economictimes.indiatimes.com/news/politics-and-nation/how-chennai-one-of-the-worlds-wettest-major-cities-ran-out-of-water/articleshow/80680182.cms.

IPCC. (2023). *Climate Change 2023: Synthesis Report a Report of the Intergovernmental Panel on Climate Change*. Retrieved from https://www.ipcc.ch/report/ar6/syr/downloads/report/IPCCAR6SYRFullVolume.pdf.

Kingdom, B., Liemberger, R., & Marin, P. (2006). *The challenge of reducing Non-Revenue Water (NRW) in developing countries how the private sector can help: a look at performance-based service contracting*. Retrieved from https://documents1.worldbank.org/curated/en/385761468330326484/pdf/394050Reducing1e0water0WSS81PUBLIC1.pdf.

London floods highlight failure to address climate change and its consequences. (2021, July 28). *World Socialist Web Site*. Retrieved from https://www.wsws.org/en/articles/2021/07/28/floo-j28.html.

Markham, D. (2025, April 28). *Californians could save $6.5 Billion with community solar & storage—CleanTechnica*. CleanTechnica. Retrieved from https://cleantechnica.com/2025/04/28/californians-could-save-6-5-billion-with-community-solar-storage/.

Martin. (2023, October 20). *Cities—United Nations Sustainable Development Action 2015*. United Nations Sustainable Development. Retrieved from https://www.un.org/sustainabledevelopment/cities/?ysclid=mbqwjl2jjh156447569.

Mitlin, D. (2019). *Why is water still unaffordable for sub-Saharan Africa's urban poor? International Institute for Environment and Development.* Retrieved from https://www.iied.org/17353iied.

Roosevelt Island Tidal Energy Project, New York. (n.d.). *Power Technology.* Retrieved from https://www.power-technology.com/projects/roosevelt-island-tidal-energy-project-new-york/?cf-view.

Short-Term Energy Outlook Supplement: Drought Effects on California Electricity Generation and Western Power Markets. (2022). Retrieved from https://www.eia.gov/outlooks/steo/special/supplements/2022/2022sp02.pdf.

Team, I. (2025, June 7). *How does urban sprawl affect the environment? The Institute for environmental research and education.* The Institute for Environmental Research and Education. Retrieved from https://iere.org/how-does-urban-sprawl-affect-the-environment/.

UNESCO. (2020, April 1). *The United Nations world water development report 2020: water and climate change.* Unesdoc.unesco. Retrieved from https://unesdoc.unesco.org/ark:/48223/pf0000372985.

United Nations. (2018). World urbanization prospects: The 2018 revision. *Population and Development Review, 24*(4), 883. https://doi.org/10.2307/2808041

USGBC. (2020). Usgbc.org. Retrieved from https://www.usgbc.org/projects/city-greensboro-nc.

Usgbc.org. Retrieved from https://build.usgbc.org/l/413862/2023-07-28/246vh9s/413862/16905657706biUEe1d/LEEDv41LFCExistingCitiesBetaGuideJuly2023.pdf.

Widespread water shut-offs in US city of Detroit prompt outcry from UN rights experts. (2014, June 25). *UN News.* Retrieved from https://news.un.org/en/story/2014/06/471652.

World Bank—Cities Key to Solving Climate Crisis. (2023, May 18). *World Bank.* Retrieved from https://www.worldbank.org/en/news/press-release/2023/05/18/cities-key-to-solving-climate-crisis.

9

Industrial and Agricultural Water Conservation and Energy

Robert C. Brears

Abstract This chapter examines practical water and energy efficiency strategies for the industrial and agricultural sectors, with a focus on technologies that reduce resource consumption and enhance operational resilience. It examines key practices, including water audits, submetering, cooling tower controls, water reuse, soil moisture monitoring, rainwater harvesting, and irrigation scheduling. Emphasizing the water-energy nexus, the chapter highlights how integrated solutions can lower costs, enhance sustainability, and respond to climate and regulatory pressures. It also outlines enabling policies and programs that support adoption. This comprehensive overview provides decision-makers with actionable insights to advance efficient and sustainable water and energy use in both sectors.

Keywords Industrial water efficiency · Agricultural water conservation · Sustainable water management · Energy-efficient irrigation · Water-energy nexus solutions

© The Author(s), under exclusive license to Springer Nature
Switzerland AG 2025
Y. Ermolaeva et al., *Rethinking Water and Energy for a Sustainable Future*, Palgrave
Studies in Climate Resilient Societies, https://doi.org/10.1007/978-3-032-04485-3_9

Introduction

Water and energy are inextricably linked across industrial and agricultural systems. Using water more efficiently reduces the energy required for extraction, treatment, distribution, and heating. Similarly, energy-efficient technologies and operations often support more sustainable water use. As global demand for resources grows, improving efficiency is crucial for overcoming resource constraints, mitigating environmental impacts, and enhancing operational resilience.

Industrial and agricultural sectors together account for a significant share of global water withdrawals. In the industry, water is essential for cooling, processing, cleaning, and sanitation, with usage patterns varying by activity. Agriculture remains the world's largest water user, primarily for irrigation purposes. Both sectors face intensifying pressures from water scarcity, rising energy costs, climate impacts, and regulatory demands. Meeting these challenges requires integrated water–energy management strategies that are both scalable and adaptable.

This chapter provides a practical guide to improving water and energy efficiency in industrial and agricultural settings. It outlines technologies, practices, and enabling conditions that support the transition to resource-efficient operations. For industry, the chapter explores facility water audits, submetering, advanced cooling tower controls, and internal water reuse. These strategies help reduce withdrawals, lower chemical use, and optimize energy performance. For agriculture, the chapter covers drip irrigation, soil moisture monitoring, rainwater harvesting, and farm wastewater recycling, all of which support smarter water use and reduced energy demand for pumping and application.

Rather than focusing on individual business case studies, the chapter emphasizes the role of public sector interventions in enabling broad adoption. Municipal and regional programs, including conservation-linked rate structures, performance-based rebates, shared reuse infrastructure, and digital management platforms, create the governance conditions needed for effective scale-up.

By pairing technical solutions with institutional mechanisms, this chapter offers a holistic view of how integrated water–energy efficiency can be achieved. It serves as a practical reference for facility managers, policymakers, and planners seeking to implement sustainability strategies across diverse systems. Ultimately, it underscores that lasting water and energy conservation is not just a matter of technology, but also of governance, coordination, and long-term investment.

Industrial Water Conservation

Industrial water conservation is a key component of sustainable resource management, enabling facilities to reduce water consumption, lower energy demand, and improve operational efficiency. Industrial processes, ranging from cooling and processing to sanitation and equipment cleaning, not only consume large volumes of water but also require substantial energy to pump, heat, or treat it, thereby compounding resource pressures. With growing water scarcity, rising utility costs, and stricter environmental regulations, industries increasingly recognize the need to adopt integrated water efficiency strategies.

This section outlines practical solutions for enhancing water use efficiency in industrial settings. It covers structured water audits to identify system inefficiencies, submetering to monitor usage patterns, advanced cooling tower controls to minimize evaporation and blowdown, and internal water reuse systems that reduce reliance on potable water sources. These measures help reduce operational costs, mitigate environmental impacts, and contribute to broader corporate sustainability goals.

To demonstrate how enabling policy environments can support industrial conservation, this section includes an example of a municipal program that incentivizes water efficiency through performance-based eligibility and compliance requirements. Such initiatives show how cities and regions can facilitate the uptake of conservation practices by aligning financial incentives with sustainable operations.

By pairing technical practices with supportive governance mechanisms, the section emphasizes that effective water conservation is not solely a matter of technology adoption, but also requires institutional coordination, planning, and a long-term commitment.

Advanced Cooling Tower Controls

Advanced cooling tower controls offer significant potential for improving water efficiency in industrial operations. These systems manage the concentration of dissolved minerals in recirculating water by continuously monitoring and adjusting conductivity levels. The primary objective is to maintain optimal cycles of concentration, which refers to the ratio of dissolved solids in recirculating water compared to fresh make-up water. Operating at higher cycles reduces the need for blowdown. It lowers make-up water demand, directly conserving water while decreasing the energy required to treat, pump, and cool additional water volumes.

By automatically adjusting chemical dosing based on real-time measurements, advanced controllers help maintain cleaner heat exchange surfaces. This reduces scaling and fouling, improving heat transfer efficiency and cutting the energy required for cooling processes. Dynamic controls ensure the system adapts to fluctuations in water chemistry and operational loads, thereby maintaining optimal performance across a wide range of conditions. Avoiding overfeeding of treatment chemicals not only saves money but also lowers environmental discharge volumes and treatment needs, supporting both sustainability and compliance.

Multiple variations of these controllers are available from different manufacturers and are accessible for widespread deployment. They can be integrated into newly designed cooling tower systems or retrofitted onto existing infrastructure. This flexibility supports their rapid adoption in industrial water efficiency programs. Effective implementation requires coordination with a water treatment professional to calibrate system parameters based on site-specific needs, such as local water hardness, corrosion risk, and operational load profiles.

In water-stressed or energy-constrained environments, these systems offer dual benefits: reduced freshwater intake and lower energy use for cooling, treatment, and pumping. Their real-time data capabilities also facilitate remote monitoring, predictive maintenance, and better system diagnostics, enabling further operational efficiencies and long-term cost savings.

Advanced cooling tower controls are particularly effective in facilities with high cooling loads, such as power plants, refineries, data centers, and manufacturing complexes. Their ability to maintain optimal system conditions through automated monitoring and feedback makes them a powerful solution for reducing water consumption, mitigating energy demand, and improving chemical management. Their ease of integration with both new and legacy infrastructure adds further value, making them a scalable, cost-effective measure in industrial water conservation strategies.

Overall, advanced cooling tower controls support a holistic approach to resource efficiency, delivering measurable water savings, improved energy performance, and reduced chemical use in a single integrated system.[1,2]

Industrial Water Reuse

Industrial water reuse refers to the practice of recycling treated water for industrial purposes, thereby reducing dependence on freshwater sources and alleviating the energy demand associated with water extraction, treatment, and discharge. This approach includes the use of reclaimed municipal wastewater or onsite industrial wastewater streams in processes such as cooling, boiler feed, and general washing. Water reused in industrial settings typically has limited human contact, making it less costly and less energy-intensive to treat to fit-for-purpose quality standards.

Water reuse can take two primary forms: internal recycling and external reuse. Internal recycling involves treating wastewater onsite for reuse within the same process, often forming a closed loop. This significantly reduces overall water intake and the energy required for pumping and treatment, though complete recycling is limited by the accumulation of contaminants that are difficult to remove. External reuse, by contrast, involves treating wastewater for use in different applications within or outside the facility. For example, wastewater from one part of a manufacturing process might be treated and reused for cooling or cleaning elsewhere, displacing additional treatment and delivery energy burdens.

Major industrial water uses such as cooling and boiler feed water are well-suited to reuse due to their high volume and relatively low contaminant load. In cooling operations, which can account for up to two-thirds of industrial water use, reclaimed water can often replace freshwater with minimal treatment. Boiler feed water, which requires higher purity, may necessitate advanced treatment methods such as ultrafiltration, reverse osmosis, and ion exchange to meet stringent specifications, particularly in sectors like petrochemicals or semiconductors. Although high-grade treatment involves energy inputs, the overall energy footprint is often lower than sourcing and distributing fresh water.

The treatment level required depends on the water source and intended use. Fit-for-purpose treatment ensures that water is treated only to the extent necessary for its subsequent use, minimizing avoidable energy and chemical consumption. Technologies such as membrane filtration, membrane bioreactors, and advanced oxidation processes are increasingly adopted for their ability to produce high-quality recycled water while also reducing operational energy intensity.

Successful reuse projects rely on several enabling factors. These include proper system design, often based on pilot testing, skilled operation and maintenance, and effective concentrate management, especially when dealing with membrane-based systems. Financial viability is also key, with public–private partnerships often used to support implementation in capital-constrained regions.

As water scarcity, regulatory pressures, and energy costs rise, industrial water reuse is becoming essential to resource-smart management. It supports cost efficiency, environmental performance, and reduces energy-related emissions, making it a cornerstone of sustainable industrial development.[3,4]

Water Use Monitoring in Industrial Facilities

Water use monitoring is a foundational component of industrial water management, providing the necessary data to guide efficiency improvements, detect operational issues, and ensure regulatory compliance. Accurate measurement of water usage across systems and processes helps industrial operators identify high-consumption activities, detect leaks, and support data-driven decision-making that reduces both water and energy waste.

Metering refers to the use of utility-owned water meters that measure the total volume of water supplied to a facility. While useful for billing and broad consumption tracking, these meters often provide limited granularity, particularly when based on monthly or quarterly readings. To address this limitation, facilities increasingly turn to submetering, which involves facility-owned equipment that monitors water use in specific processes or areas, such as cooling towers, boilers, or manufacturing lines. By isolating energy-intensive systems, submetering supports not only water conservation but also reductions in associated energy use for pumping, heating, and treatment.

Advanced metering infrastructure (AMI) represents the next generation of water monitoring. These systems integrate smart meters, sensors, telemetry units, and data management software to collect and transmit usage data at hourly or even more frequent intervals. The real-time data generated through AMI provides facility managers with immediate insight into consumption patterns, enabling the prompt detection of leaks, equipment failures, or unexpected usage spikes. Many utilities now offer customer portals that allow facility personnel to access this data directly, often with tools for setting usage thresholds and receiving alerts for anomalies. Given the high energy requirements of water supply systems, AMI contributes to operational efficiency by minimizing avoidable water movement and processing costs.

In industrial settings, AMI is particularly valuable for high-volume processes where unplanned water loss can translate into significant operational disruptions or cost increases. For example, continuous monitoring of water-cooled systems or washdown processes helps maintain operational stability and avoid damage from leaks or equipment malfunction. Submeters and AMI systems also support efforts to benchmark usage across multiple facilities, allocate costs by department or tenant, and verify the outcomes of water efficiency upgrades. By detecting inefficiencies early, these systems prevent energy-intensive overuse of water resources and reduce the strain on upstream infrastructure.

Selecting the appropriate metering technology is crucial. Conventional mechanical meters may require retrofitting with reed switches to enable pulse output for data loggers. In contrast, electromagnetic and ultrasonic meters offer digital outputs and are compatible with automated data collection systems. Facilities should ensure meters are appropriately sized, as undersized or oversized meters can lead to inaccurate readings and operational inefficiencies.

By implementing submetering and leveraging AMI, industrial facilities gain the detailed visibility needed to optimize water use, reduce costs, and lower the energy footprint of water operations.[5,6]

Water Audits

Conducting a water audit is a foundational step in developing a successful water conservation program. It enables organizations to understand precisely how water is used throughout their facilities, identify areas of inefficiency, and quantify potential savings opportunities. The process begins with preparation and information gathering, including collecting facility diagrams, plumbing schematics, utility records, and an inventory of all water-using equipment. This data helps establish a baseline for water use and highlights trends in consumption that may also indicate energy waste linked to water-intensive processes.

Following data collection, a detailed physical survey is conducted. This involves walking through the facility with operations staff to observe how and where water is used, measuring flow rates, identifying equipment and processes that consume water, and checking for leaks or inefficiencies. Because water movement, treatment, and heating are energy-intensive, inefficiencies in water use frequently correlate with high energy demand. Water meters are calibrated, and where permanent meters are not installed, temporary meters or manual measurements are used to assess usage. It is also important to test water quality, as this can influence decisions about reuse within the facility. Exterior water use, including irrigation, is also reviewed, with a focus on sprinkler system efficiency and water scheduling.

After the physical inspection, the audit findings are compiled into a report. This document includes updated diagrams, actual versus recommended flow rates, operational schedules, water flow charts, and cost evaluations. The report forms the basis for identifying areas with the most significant potential for water savings and supports the development of a conservation plan. Critically, the audit helps pinpoint opportunities where reducing water use can also cut energy costs, particularly in systems such as cooling towers, boilers, or hot water lines. Additionally, the audit helps uncover hidden issues, such as underground leaks, which may not be evident without detailed measurement and comparison against utility records.

Because water-intensive processes often require energy for pumping, heating, or treatment, addressing inefficiencies through audits directly contributes to energy reduction. Facilities that conduct thorough audits often realize significant cost savings across their utility bills, while also enhancing operational resilience.

Publicizing the success of water conservation initiatives serves multiple purposes. Internally, it reinforces the organization's commitment and helps maintain employee engagement. Staff should be kept informed through regular updates, newsletters, and visual displays that highlight the savings achieved. Acknowledging individual and team contributions can further motivate participation and encourage a culture of conservation.

Externally, sharing success stories can enhance the organization's reputation as a responsible steward of natural resources. Communicating quantifiable achievements, such as gallons and kilowatt-hours saved, can enhance credibility, generate positive media coverage, and position the organization as a leader in sustainability.[7]

Case Study: Enabling Industrial Water Efficiency Through the City of Toronto's Industrial Water Rate Program

The City of Toronto promotes industrial water conservation through a performance-based policy instrument known as the Industrial Water

Rate Program. This initiative offers manufacturers a reduced water rate, known as the Block 2 rate, to encourage the adoption of water-efficient technologies and data-driven management practices. The program links financial incentives with conservation outcomes, supporting both economic competitiveness and responsible resource use. By decreasing water consumption, the program also helps reduce the energy demands associated with water intake, treatment, and discharge, contributing to broader efficiency and emissions reduction goals.

Eligible businesses must use more than 5,000 cubic meters of water annually, fall under the industrial property tax class, and comply fully with the City's Sewers By-law. Participation requires submission of a comprehensive water conservation plan, which Toronto Water must approve. The plan typically outlines metering strategies, leak detection systems, process improvements, and opportunities for reuse. Such measures not only conserve water but also reduce the energy required for pumping, heating, and treating process water. Once approved, the lower rate applies retroactively to the submission date, allowing for near-term financial returns on investments in efficiency measures.

The rate structure applies the general Block 1 rate to the first 5,000 cubic meters, with a 30 percent discount on usage exceeding this amount. In 2025, this translates to a rate of $4.6872 per cubic meter for the first tier and $3.2809 per cubic meter for additional consumption. For an industrial facility using 100,000 cubic meters annually, the savings can exceed $130,000. These cost savings can be reinvested in energy-efficient equipment upgrades, smart water monitoring, and closed-loop reuse systems, amplifying the program's resource efficiency benefits.

To maintain eligibility, participants must submit annual progress reports detailing updates to conservation technologies and operational performance. Facilities must also stay within prescribed discharge limits. Compliance is closely monitored, with a tiered response to Notices of Violation ranging from technical consultations to potential removal from the program. Re-entry requires either demonstrated resolution or sustained compliance for at least 12 months.

The program accommodates mixed-use facilities, provided they can isolate industrial water flows using submeters. Where physical submetering is infeasible, businesses must submit technical documentation to justify an exemption and propose alternative verification measures.

By aligning rate structures with technology adoption and performance reporting, Toronto's Industrial Water Rate Program exemplifies how local governments can drive water efficiency through structured incentives. It also demonstrates how targeted water conservation policy can yield energy savings across industrial systems, reducing the environmental footprint of manufacturing operations. It highlights the importance of integrated monitoring and regulatory alignment in promoting conservation within industrial systems.[8]

Agricultural Water Conservation

Agricultural water conservation is crucial for alleviating the increasing pressure on limited water and energy resources. Water delivery, pumping, and treatment in agriculture are often energy-intensive, meaning inefficiencies in water use directly translate to higher energy demands and increased operational costs. By improving how and when water is applied, farms can reduce energy used for pumping, protect soil health, and boost crop resilience to changing climatic conditions. Key strategies, including drip irrigation, rainwater harvesting, soil moisture monitoring, irrigation scheduling, and water recycling, enhance overall efficiency, reduce runoff, and limit nutrient loss. These technologies not only cut reliance on freshwater and fossil energy but also contribute to stable yields and long-term agricultural viability.

This section outlines practical, technology-based approaches that form a cohesive and sustainable framework for agricultural water management. To demonstrate how enabling policy and funding environments can accelerate the adoption of such technologies, the section includes an example of a regional grant program that promotes the uptake of digital monitoring tools and water-saving infrastructure. This highlights the role of public investment in scaling resource-efficient practices across diverse farming systems.

Irrigation Scheduling

Irrigation scheduling is the process of determining when and how much water to apply to a field, based on the crop's water needs, soil characteristics, and environmental conditions. The goal is to optimize plant growth while conserving water and minimizing adverse environmental impacts. Poor irrigation scheduling can lead to under- or overwatering, both of which reduce yield and quality. Underwatering causes plant stress and reduced photosynthesis, while overwatering leads to nutrient leaching, waterlogging, and wasted energy.

Effective scheduling considers three critical factors: crop water demand, soil water availability, and the soil's storage capacity. Water use by plants varies depending on the species, growth stage, and atmospheric conditions. Soil acts as a temporary water reservoir, and its effectiveness depends on texture and structure. Water availability is determined by the amount of water remaining in the root zone between field capacity, the point at which excess water has drained, and the permanent wilting point, at which plants can no longer extract water.

The difference between field capacity and permanent wilting point defines the available water-holding capacity (AWC). Coarse-textured soils like sand have low AWC due to large pores and quick drainage, requiring frequent irrigation in small amounts. Fine-textured soils, such as clay, hold more water and allow for less frequent, yet deeper, applications. Rooting depth further determines how much of this stored water is accessible to crops. Shallow-rooted plants such as strawberries or lettuce require more frequent irrigation than deep-rooted crops like corn or alfalfa.

A key parameter in scheduling is the management allowable depletion (MAD), which defines the percentage of AWC that can be used before irrigation is necessary. While 50% depletion is a common threshold, this value can vary based on crop type, growth stage, and system capacity. For example, if a soil has 3.5 inches of available water and the MAD is set at 30%, irrigation should occur after 1.05 inches of water are used (0.3 × 3.5).

Irrigation scheduling is also affected by crop development. Early in the season, when root systems are shallow, irrigation depth should be limited.

As roots deepen, more water can be stored and applied at longer intervals. Most irrigation systems aim to refill the root zone to field capacity while leaving room for potential rainfall. In areas where rainfall is variable, scheduling may need to adapt quickly to avoid plant stress.

Multiple tools assist in irrigation scheduling. Soil moisture monitoring technologies, including tensiometers, gypsum blocks, capacitance probes, and TDR sensors, provide real-time feedback on soil conditions. Soil water balance models estimate changes in soil moisture based on evapotranspiration and precipitation. Some crops exhibit visible signs of water stress, such as color changes or wilting, but many do not, making sensor-based tools essential for informed decision-making.

The frequency and volume of irrigation are also influenced by the irrigation method used. Surface irrigation methods, such as furrows or basins, can infiltrate larger volumes, but may be limited by application uniformity. Pressurized systems, such as drip or sprinkler irrigation, provide more precise control, allowing for tailored applications based on soil and crop needs.

Scheduling decisions must also take into account seasonal patterns. Spring irrigation may be necessary for germination and early growth, while summer often requires more frequent watering due to higher evapotranspiration rates. Autumn irrigation can improve soil moisture before planting or dormancy. Winter irrigation is possible in regions without freezing temperatures and can support deep moisture storage.[9,10,11,12,13]

Rainwater Harvesting

Rainwater harvesting for agriculture is the practice of collecting, conveying, storing, and utilizing rainwater runoff for productive on-farm use, primarily in crop production. Unlike municipal or household systems, agricultural applications must account for local variability in climate, hydrology, and soil characteristics that influence crop water demand. These include infiltration rates, evapotranspiration levels, rooting depth, and rainfall distribution patterns across the season.

At its core, rainwater harvesting captures and stores water that would otherwise be lost as runoff. Rainfall from rooftops, greenhouses, roads, or sloped terrain is diverted to storage systems, such as tanks, cisterns, or ponds, for later use during dry spells or irregular rainfall periods. This provides a supplemental water source that relies on energy-efficient groundwater pumping or local water infrastructure, thereby reducing overall pressure on freshwater and energy resources.

A critical component in system design is estimating "design rainfall", the volume of rain that can be reliably expected in a typical year, usually based on a 90% probability threshold. This figure, derived from long-term meteorological records, guides decisions on catchment area size, storage capacity, and system yield. Not all rainfall becomes runoff, as factors such as surface texture, slope, infiltration, and evaporation affect the collection efficiency. The smoother and less permeable the surface, the more efficient the water capture, making material choice key in system performance.

Storage solutions vary by context. Smallholders may rely on above-ground polyethylene tanks or concrete cisterns, while larger operations might install geomembrane-lined ponds or corrugated steel silos. Effective storage systems require sediment filters, roof washers, or sand media filters to maintain water quality and prevent clogging. Overflow structures are also crucial in managing heavy rainfall and protecting the surrounding land from erosion and flooding.

In greenhouse agriculture, connected gutter systems can harvest substantial volumes of clean rainwater. For example, a single inch of rain on a one-acre greenhouse can yield more than 27,000 gallons of water. After accounting for system losses such as evaporation and debris diversion, usable yield averages about 65%, but this still offers significant conservation benefits. Harvested water can be stored and integrated into drip or sprinkler systems, lessening the energy required to extract or transport irrigation water from distant or fossil-fueled sources.

Rainwater harvesting also enhances nutrient management. By concentrating water in specific application zones, nutrients remain in the root zone longer, reducing leaching and improving fertilizer efficiency. In semi-arid or drought-prone areas, rainwater harvesting acts as a form of "runoff farming," where runoff is redirected from external catchments into terraced or bounded fields. Though exact timing and volumes can vary, the method adds a vital layer of water resilience for climate-sensitive production.

Overall, rainwater harvesting supports climate-resilient agriculture while decreasing dependence on conventional water and energy infrastructure.[14,15,16]

Soil Moisture Monitoring

Efficient irrigation requires timely and accurate information on soil water availability. Soil moisture monitoring provides the data necessary to apply water only when and where it is needed, reducing waste, conserving energy used for water delivery, and improving crop productivity. By ensuring that irrigation matches crop demand, these technologies limit over-irrigation, prevent nutrient leaching, and enhance overall water-use efficiency.

Soil moisture sensors are typically installed at multiple depths within the root zone to track real-time changes in water content. Common technologies include tensiometers (which measure soil water tension), capacitance probes, and time-domain reflectometry sensors. These tools help producers avoid unnecessary watering by identifying when soil moisture is within optimal thresholds for crop health. When used in pressurized systems, such as drip or sprinklers, avoiding overwatering results in measurable reductions in the energy required for pumping, filtering, and distribution.

To fully benefit from moisture data, farms often combine monitoring with irrigation scheduling systems. Scheduling tools integrate sensor data with crop water use models, evapotranspiration rates, and localized weather forecasts to determine the ideal timing, frequency, and duration of irrigation. This data-driven approach allows operators to adjust irrigation in response to environmental conditions rather than relying on static schedules. The result is not only improved water-use efficiency but also reduced runtime of irrigation systems, directly cutting energy consumption.

Advanced platforms now automate this entire process. By linking sensors to cloud-based controllers, farms can remotely manage irrigation systems that apply water with high precision and accuracy. These integrated systems can also detect issues such as line blockages or leaks, reducing water loss and avoiding the added energy costs associated with inefficiencies.

In water-stressed or energy-constrained regions, these tools help decouple agricultural production from the volatility of external water and power supplies. They also support groundwater sustainability by reducing drawdown rates and avoiding excess pumping during peak energy demand periods, when electricity costs and grid stress are highest.

Selecting the right system depends on factors such as farm scale, crop type, soil characteristics, and infrastructure. Proper placement, calibration, and maintenance of sensors are essential for accurate readings. Similarly, irrigation schedules should be regularly updated to reflect seasonal changes, crop growth stages, and observed weather patterns.

Together, soil moisture monitoring and irrigation scheduling form a cornerstone of precision water management, enabling farmers to do more with less, less water, less energy, and fewer inputs overall, while protecting long-term soil and water resources. These technologies are vital components in the transition to climate-smart agriculture.[17,18,19]

Water Recycling for Irrigation

Water recycling for agricultural irrigation involves the intentional collection, treatment, and reuse of water for crop production and farm operations. As agricultural demands intensify alongside declining freshwater availability, recycling offers a practical strategy to improve water use efficiency, reduce environmental discharge, and enhance the resilience of farming systems. The process supports circularity by treating wastewater not as waste but as a recoverable resource that can re-enter the production cycle, reducing the need for energy-intensive withdrawals from surface or groundwater sources.

Sources of water suitable for recycling in agriculture include on-farm runoff, drainage water, livestock facility effluent, municipal wastewater, and some forms of lightly contaminated industrial process water. The selection of a water source depends on local availability, crop requirements, and the intended reuse application. On-farm sources such as irrigation return flows can be captured through drainage infrastructure and routed to collection basins for subsequent reuse. Livestock wastewater, once treated to remove solids and pathogens, can be applied to fields as both irrigation and a nutrient source.

The treatment level for recycled water is determined by its intended use. Primary treatment may involve sedimentation to remove solids, while secondary processes focus on removing organic matter and pathogens. In some cases, advanced treatment using membrane filtration, UV disinfection, or chemical oxidation may be necessary, particularly when the water is used for crops that are consumed raw. Constructed wetlands also offer a low-energy, land-based treatment option, particularly suited to smaller or integrated farming systems.

Recycled water can be distributed through various irrigation systems, including surface irrigation, sprinkler irrigation, and drip irrigation methods. Drip irrigation is especially energy-efficient when combined with recycled water, as it minimizes both water volume and pumping demands by delivering moisture directly to root zones. Surface methods may be appropriate for forage or non-edible crops where water quality permits. System design must consider filtration to prevent emitter clogging, especially in drip systems where fine particulates or biofilms can accumulate.

Soil texture influences how recycled water is applied. Coarse soils with low water-holding capacity benefit from smaller, more frequent applications, ensuring optimal water and nutrient uptake. Fine-textured soils retain moisture for longer periods, supporting deeper, less frequent irrigation. Application strategies are tailored to match soil characteristics, enhancing efficiency and maintaining consistent crop performance.

Crops have varying water quality needs based on their growth habits and nutrient uptake patterns. Leafy vegetables and fruiting crops benefit from high-quality recycled water, particularly when grown for raw consumption. In contrast, cereals and fodder crops perform well with a broader range of recycled water qualities. Recycled water containing plant-available nutrients, such as nitrogen and phosphorus, supports crop productivity. Root zone monitoring and soil testing enhance nutrient alignment and application efficiency.

Beyond direct field application, recycled water can support other agricultural uses, including greenhouse humidity control, facility wash-down, and equipment cleaning. In integrated systems, treated water can also be utilized in aquaponics, where fish waste contributes to nutrient loads that support plant growth in a closed-loop configuration. Surplus treated water can also be used for managed aquifer recharge, reducing energy demand during peak irrigation seasons by restoring groundwater levels when energy prices are lower.

By substituting treated water for freshwater abstraction, water recycling reduces pressure on both hydrological and energy systems, supporting more sustainable agricultural practices in resource-constrained environments.

Case Study: Technology Adoption Through the Farms of the Future Program, New South Wales, Australia

The *Farms of the Future – Agtech Grant Program* in New South Wales, Australia, promotes the integration of advanced agricultural technologies (Agtech) to enhance water efficiency, resource management, and farm resilience. Central to the program is its focus on connected sensor-based technologies that enable real-time monitoring of key on-farm variables. These technologies support more efficient irrigation and pump oper-ation, reducing both water use and the energy required to move and manage that water.

Through competitive grants of up to AUD 35,000, eligible farmers can invest in a suite of Agtech solutions from a curated catalogue. These include Internet of Things (IoT) devices such as soil moisture sensors, weather stations, flow meters, and pump controllers. Together, these technologies provide critical data on water availability, crop and soil conditions, and irrigation performance, enabling farmers to make precise and informed decisions about water application and overall farm opera-tions. Optimized irrigation scheduling helps reduce the energy intensity of farming by minimizing unnecessary pumping and ensuring water is applied only when and where needed.

A core requirement of the program is the development of an Agtech Plan, tailored to each farm's operations. Only technologies aligned with this plan are eligible for funding, ensuring that tech adoption is strategic and fit-for-purpose. These solutions are designed to support the major agricultural sectors in the region: livestock (beef and sheep), grains, cotton, and horticulture (tree crops and vines). Farmers also agree to share anonymized weather data from their IoT devices, supporting broader seasonal forecasting and data-informed policy development.

The program is tightly integrated with digital training. Farmers must complete an online Agtech training course before becoming eligible for funding. This ensures they understand how to select, install, and use the technologies effectively. The emphasis on both hardware and digital literacy is key to ensuring lasting adoption and benefits. Improved knowledge of system performance empowers farmers to implement water and energy-saving strategies that lower operating costs and reduce environmental impact.

Importantly, the grant does not fund technology retroactively or allow the use of non-approved devices. Only technologies that comply with the program's standards, particularly in terms of interoperability and reliability, are eligible. This ensures the consistency and quality of data generated across participating farms.

By linking funding to both strategic planning and training, *Farms of the Future* offers a structured pathway for scaling Agtech adoption across regional agricultural communities. The emphasis on water monitoring and management technologies reflects a broader recognition of the need for precision agriculture to address challenges of water scarcity, climate volatility, and economic pressure. Crucially, the program demonstrates how targeted investment in digital infrastructure can simultaneously drive down water consumption and farm energy demand, supporting the transition to a more resource-efficient agricultural sector.[20]

Best Practices and Conclusion

This section distills the key insights from the chapter into actionable best practices for improving water and energy efficiency across industrial and agricultural operations. While technologies and strategies vary by sector and setting, common themes emerge, including the importance of monitoring, process optimization, and fit-for-purpose reuse. By implementing proven approaches, facility managers and farm operators can reduce resource consumption, lower costs, and enhance resilience to water and energy challenges. The following best practices offer a practical summary of effective measures, followed by a concluding reflection on the broader significance of integrated water-energy management for sustainable development.

Industrial Best Practices: Water and Energy Efficiency in Industry

Effective industrial water conservation reduces pressure on both water and energy systems while supporting cost savings and operational efficiency. The following best practices reflect proven, technology-based strategies:

1. **Conduct a Comprehensive Water Use Inventory or Audit**

   Map showing where water enters, is used, and exits the facility. Identify inefficiencies and high-use areas to target conservation efforts and uncover energy-intensive processes.

2. **Submeter and Monitor High-Use Systems and Processes**

   Install submeters and flow sensors on cooling towers, boilers, and key equipment to gather detailed usage data, detect leaks early, and verify efficiency improvements.

3. **Optimize Cooling Towers with Advanced Controls**

Utilize conductivity-based controllers to manage cycles of concentration, minimize blowdown, and optimize chemical dosing, thereby reducing both water and energy consumption in cooling.

4. Improve Boiler and Steam System Efficiency

Recover blowdown heat and reuse condensate to lower water intake, energy demand, and chemical treatment requirements across steam systems.

5. Replace Single-Pass Cooling with Recirculating Systems

Eliminate once-through cooling wherever possible by installing closed-loop systems that dramatically reduce water consumption and thermal discharge.

6. Implement Internal Water Recycling and Reuse

Treat and reuse process water within operations, such as rinsing or cooling, where quality allows, thereby reducing intake volumes and associated energy costs.

7. Automate Monitoring and Leak Detection with Smart Infrastructure

Adopt advanced metering infrastructure (AMI) and real-time monitoring to track anomalies, enable predictive maintenance, and reduce operational waste.

8. Train Staff and Maintain Preventive Maintenance Programs

Educate personnel on efficient practices and establish regular inspection routines for valves, pumps, and piping to minimize leaks and hidden system losses.

Agricultural Best Practices: Water and Energy Efficiency in Agriculture

Efficient water management in agriculture is crucial for conserving resources, minimizing energy inputs, and enhancing climate resilience. The following best practices help optimize water use while maintaining soil health and agricultural productivity:

1. Use Soil Moisture Monitoring Technologies

Install tensiometers, capacitance probes, or TDR sensors to track real-time soil moisture in the root zone. This supports accurate irrigation timing, reducing overwatering and energy waste.

2. Schedule Irrigation Based on Crop, Soil, and Weather Data

Use evapotranspiration rates, crop growth stages, and soil characteristics to calculate irrigation timing and volume, reducing unnecessary water and pump energy use.

3. Match Irrigation Frequency and Depth to Soil Texture and Rooting Depth

Tailor irrigation to specific soil types (e.g., sandy vs. clay) and crop root zones to improve moisture retention and minimize deep percolation or runoff losses.

4. Implement On-Farm Water Recycling Systems

Collect and treat on-farm runoff, drainage, or livestock wastewater for reuse in irrigation, reducing freshwater withdrawals and energy used in pumping.

5. Harvest and Store Rainwater for Productive Use

Capture rainfall from rooftops or impermeable surfaces in tanks or ponds to provide supplementary irrigation and lessen reliance on energy-intensive groundwater extraction.

6. Select Irrigation Methods Based on Efficiency and Crop Needs

Utilize drip, sprinkler, or surface irrigation systems strategically, based on crop sensitivity, water quality, and soil conditions, to minimize water and energy consumption.

7. Automate Irrigation with Integrated Monitoring and Scheduling Tools

Combine soil sensors, weather forecasts, and scheduling software to control irrigation systems automatically, reducing human error and optimizing pump runtime.

8. Align Irrigation Scheduling with Management Allowable Depletion (MAD)

Irrigate based on how much of the soil's available water has been depleted, using thresholds tied to crop type and growth stage to avoid plant stress or overuse.

Conclusion

Water and energy efficiency are fundamental to building resilient industrial and agricultural systems in an era of climate stress, regulatory pressure, and rising operational costs. As this chapter has demonstrated, a diverse range of technologies and practices, from advanced monitoring to precision reuse, can help facilities and farms reduce consumption, enhance performance, and achieve long-term sustainability goals. However, technological upgrades alone are insufficient. Success depends on aligning these interventions with supportive policy frameworks, capacity building, and investment mechanisms that enable broad adoption and scale.

Integrated management of water and energy yields multiple co-benefits: reduced operational expenses, enhanced compliance, improved environmental outcomes, and greater adaptability to changing resource conditions. Industrial and agricultural actors must therefore prioritize system-level thinking, embedding efficiency into both infrastructure and daily operations. Public-sector support through pricing reforms, grant programs, and regulatory incentives can accelerate uptake, ensuring that proven practices move from pilot initiatives to sector-wide norms.

Ultimately, effective water and energy conservation is a shared responsibility that requires collaboration across sectors, levels of governance, and technical disciplines. By treating conservation as both a technical and institutional challenge, stakeholders can develop solutions that are not only cost-effective but also equitable and environmentally sound, contributing to a more sustainable and resilient future.

Notes

1. US Department of Energy, "Water-Efficient Technology Opportunity: Advanced Cooling Tower Controls," https://www.energy.gov/femp/water-efficient-technology-opportunity-advanced-cooling-tower-controls.
2. Robert C. Brears, *Water Resources Management: Innovative and Green Solutions*, 2 ed. (Berlin, Boston: De Gruyter, 2024).
3. Josef Lahnsteiner, Patrick Andrade, and Rajiv D. Mittal, "Industrial Water Reuse and Recycling, an Introduction," in *Handbook of Water and Used Water Purification*, ed. Josef Lahnsteiner (Cham: Springer International Publishing, 2024).
4. US EPA, "Basic Information About Water Reuse," https://www.epa.gov/waterreuse/basic-information-about-water-reuse.
5. R.C. Brears, *Regional Water Security* (Wiley, 2021).
6. Brears, *Water Resources Management: Innovative and Green Solutions*.

7. US EPA, "A Water Conservation Guide for Commercial, Institutional and Industrial Users," (1999), https://www.epa.gov/sites/default/files/2021-01/documents/cii-users-guide.pdf.

8. Toronto Water, "Industrial Water Rate Program," https://www.toronto.ca/services-payments/water-environment/how-to-use-less-water/water-efficiency-for-business/industrial-water-rate-program/.

9. Oregon State University, "Describe Basic Principles of Scheduling Irrigation for Efficient Use of Water Resources.," https://forages.oregonstate.edu/nfgc/eo/onlineforagecurriculum/instructormaterials/availabletopics/irrigation/scheduling.

10. FAO, "Irrigation Water Management: Irrigation Scheduling," (1989), https://www.fao.org/4/t7202e/t7202e00.htm#Contents.

11. University of Minnesota, "Basics of Irrigation Scheduling," https://extension.umn.edu/irrigation/basics-irrigation-scheduling.

12. SAI Platform, "Irrigation Scheduling," (2010), https://www.saiplatform.org/uploads/Library/Technical%20Brief%206.%20Irrigation%20Scheduling.pdf#page4.

13. Robert C. Brears, *Sustainable Water-Food Nexus*, Circular Economy, Water Management, Sustainable Agriculture (De Gruyter, 2025).

14. FAO, "Compendium on Rainwater Harvesting for Agriculture in the Caribbean Sub-Region: Concepts, Calculations and Definitions for Small, Rain-Fed Farm Systems," (2014), https://openknowledge.fao.org/server/api/core/bitstreams/9e1e6970-26d5-4d07-94ce-7202552e4988/content#page10.

15. University of Massachusetts Amherst, "Rainwater Harvesting," https://www.umass.edu/agriculture-food-environment/greenhouse-floriculture/fact-sheets/rainwater-harvesting.

16. R.C. Brears, *Nature-Based Solutions to 21st Century Challenges* (Oxfordshire, UK: Routledge, 2020).

17. Agriculture Victoria, "Soil Moisture Monitoring," https://agriculture.vic.gov.au/__data/assets/pdf_file/0006/577023/Soil-Moisture-Monitoring-fact-sheet-Dec-2017.pdf.

18. University of Wyoming Extension, "Methods and Techniques for Soil Moisture Monitoring," (2018), https://wyoextension.org/pub lications/html/B1331/.

19. University of Minnesota, "Soil Moisture Sensors for Irrigation Scheduling," https://extension.umn.edu/irrigation/soil-moisture-sensors-irrigation-scheduling#pros%2C-cons-and-costs-of-soil-water-tension-sensors-1751861.

20. NSW Government, "Farms of the Future—Agtech Grant Program," https://www.nsw.gov.au/grants-and-funding/farms-of-future-grant-program#:~:text=The%20Farms%20of%20the%20Future,sectors%3A%20Cotton%2C%20Livestock%20(sheep.

References

Agriculture Victoria. *Soil moisture monitoring*. Retrieved from https://agricu lture.vic.gov.au/__data/assets/pdf_file/0006/577023/Soil-Moisture-Monito ring-fact-sheet-Dec-2017.pdf.

Brears, R. C. (2020). *Nature-based solutions to 21st century challenges*. Rout-ledge.

Brears, R. C. (2021). *Regional water security*. Wiley.

Brears, R. C. (2024). *Water resources management: Innovative and green solutions* (2nd ed.). De Gruyter. https://doi.org/10.1515/9783111028101

Brears, R. C. (2025). *Sustainable water-food nexus*. De Gruyter. https://doi.org/10.1515/9783111341385

FAO. (1989). *Irrigation water management: Irrigation scheduling*. Retrieved from https://www.fao.org/4/t7202e/t7202e00.htm#Contents.

FAO. (2014). *Compendium on rainwater harvesting for agriculture in the caribbean sub-region: Concepts, calculations and definitions for small, rain-fed farm systems*. Retrieved from https://openknowledge.fao.org/server/api/core/bitstreams/9e1e6970-26d5-4d07-94ce-7202552e4988/content#page10.

Lahnsteiner, J., Andrade, P., & Mittal, R. D. (2024). Industrial water reuse and recycling, an introduction. In J. Lahnsteiner (Ed.), *Handbook of water and used water purification* (pp. 1095–1129). Springer International Publishing.

NSW Government. *Farms of the Future—Agtech Grant Program*. Retrieved from https://www.nsw.gov.au/grants-and-funding/farms-of-future-grant-pro gram#:~:text=The%20Farms%20of%20the%20Future,sectors%3A%20C otton%2C%20Livestock%20(sheep.

Oregon State University. *Describe basic principles of scheduling irrigation for efficient use of water resources*. Retrieved from https://forages.oregonstate. edu/nfgc/eo/onlineforagecurriculum/instructormaterials/availabletopics/irr igation/scheduling.

SAI Platform. (2010). *Irrigation scheduling*. Retrieved from https://www.saipla tform.org/uploads/Library/Technical%20Brief%206.%20Irrigation%20S cheduling.pdf#page4.

Toronto Water. *Industrial Water Rate Program*. Retrieved from https://www.tor onto.ca/services-payments/water-environment/how-to-use-less-water/water-efficiency-for-business/industrial-water-rate-program/.

University of Massachusetts Amherst. *Rainwater harvesting*. Retrieved from https://www.umass.edu/agriculture-food-environment/greenhouse-flo riculture/fact-sheets/rainwater-harvesting.

University of Minnesota. *Basics of irrigation scheduling*. Retrieved from https:// extension.umn.edu/irrigation/basics-irrigation-scheduling.

University of Minnesota. *Soil moisture sensors for irrigation scheduling*. Retrieved from https://extension.umn.edu/irrigation/soil-moisture-sensors-irrigation-scheduling#pros%2C-cons-and-costs-of-soil-water-tension-sensors-175 1861.

University of Wyoming Extension. (2018). *Methods and techniques for soil moisture monitoring*. Retrieved from https://wyoextension.org/publications/ html/B1331/.

US Department of Energy. *Water-efficient technology opportunity: Advanced cooling tower controls*. Retrieved from https://www.energy.gov/femp/water-efficient-technology-opportunity-advanced-cooling-tower-controls.

US EPA. *Basic information about water reuse*. Retrieved from https://www.epa. gov/waterreuse/basic-information-about-water-reuse.

US EPA. (1999). *A water conservation guide for commercial, institutional and industrial users*. Retrieved from https://www.epa.gov/sites/default/files/2021-01/documents/cii-users-guide.pdf.

US EPA. *Water reuse for industrial applications resources*. Retrieved from https:// www.epa.gov/waterreuse/water-reuse-industrial-applications-resources.

Index

Zeitfracht Medien GmbH
Ferdinand-Jühlke-Straße 7
99095 Erfurt, Deutschland
produktsicherheit@kolibri360.de